杯子蛋糕幸福上桌

糕體、糖霜到裝飾，輕鬆完成

戚黛黛、蒙順意 著

作者序

從小到大我就很喜歡烘焙，不論是餅乾還是蛋糕，我都樂於嘗試。像很多小朋友一樣，小的時候我就喜歡扮演大人的角色，照顧別人的起居飲食，最直接的方法就是做一些甜點給大家吃。看到家人朋友品嘗甜點的幸福模樣，我總是非常開心！長大以後，幸福對我而言，就是可以飛到世界各地品嘗美食，了解每個地方獨有的飲食文化和烹飪方法。為了製作出更健康美味的甜點，我跟隨不少老師學習烘焙，並完成了HKUSPACE為期8個月的營養學課程。

旅遊和烘焙是我人生中的兩大樂趣，我非常樂意將這份快樂與大家分享，於是就有了這本《杯子蛋糕幸福上桌》！在這本書中，我選擇了12個我去過並留有深刻印象的地方，運用當地的美食或獨特的材料，創造出我心目中代表這個地區的cupcake。每個地方和每個cupcake我都寫了一段文字，介紹它們對我的意義或這個蛋糕的特別之處。我還邀請了和我一樣喜愛做蛋糕、對蛋糕裝飾頗有經驗與心得的好友Monica，與我一起研究cupcake的味道和裝飾。

我希望透過這本書，描繪出一張世界美食地圖，讓大家足不出戶，也可以跟隨不同口味的cupcake感受不同地方的飲食文化！我更希望大家能帶著輕鬆的心態跟著我和Monica一起動手做cupcake！你會發現，即使用基礎的材料和簡單的方法，只要花多一點心思去做裝飾，也可以做出讓人賞心悅目的cupcake！

戚黛黛

多年前，在一個偶然機會下與朋友一起去學做蛋糕，便不能自拔地迷戀上了這個甜點陷阱，自此對烘焙蛋糕產生了濃厚興趣。我慢慢地由基本蛋糕製作開始，一路跟隨不同資深甜點師學習，並不斷與朋友分享心得和成果。

最開心的是在朋友生日時送上獨一無二的自製蛋糕，朋友收到蛋糕時的喜悅給了我很大鼓舞，讓曾當過時裝設計師的我重拾創作的快樂！為了提升自己，做出更具設計感的蛋糕，我完成了美國著名烘焙學校惠爾通（Wilton School）的蛋糕裝飾技巧課程以及英國PME糖藝專業文憑課程。在與不同國家糖藝師學習的過程中，我深深地體會到蛋糕裝飾的魅力！

糖藝師就好像魔術師，將原本平淡無奇的蛋糕變成藝術品。而好的蛋糕不僅要求美味，也講究造型、構思、色彩和對主題的呼應，蛋糕裝飾的環節更可充分發揮個人創意，給創作者帶來莫大的滿足。

已經取得惠爾通糖藝教師資格的我，希望將所學到的蛋糕裝飾技巧跟讀者們一起分享。只要有興趣，你也可以動手製作一個專業級的精緻蛋糕！

蒙順意

推薦序

由參選香港小姐到現在，我與黛黛相識已超過10年。過往經常到她家中做客的我，有幸一嘗她非凡的廚藝，由她親手製作的甜點尤其不能錯過。

黛黛對創製甜點的熱忱多年來有增無減：她會特意到世界各地找尋關於甜點的食譜或靈感；在友人的聚會上，她會親自製作各式各樣主題獨特的甜點，比如生日蛋糕、cupcake、餅乾等等，為在場的朋友送上一份甜蜜蜜的祝福。她也常常在互聯網上分享新菜式，除了賣相精緻外，味道亦令人一試難忘。得知她要出書，我當然替她高興。書裡除了分享做甜點的心得、富創意的食譜外，還有她在世界各地尋食覓味的經歷！最有趣的是，當你跟著這本書做甜點，感覺就好像跟著黛黛一起周遊列國、尋找美食一樣！ Don't forget to bring your passport！

楊洛婷／2003年香港小姐亞軍及著名活動司儀

當你感到情緒低落、心情鬱悶的時候，你會選擇用什麼方法去讓自己開心？有些人會瘋狂購物，有些會看書、看海、聽音樂、睡覺、洗澡、唱歌、漫無目的四處逛……總之各適其適。

但我相信有一種方法應該是最多人用，就是吃，尤其是甜點或者蛋糕。當你見到一個精美的蛋糕放在眼前，還未品嘗，那種賞心悅目的感覺就已經將你的開心指數上升了一半；接著將一小口放入口中，從舌尖一直刺激你的感官、神經，頓時就會使你變得開懷。為何簡單的材料就有這神奇的效果？有沒有想過要感謝為你製造出這個蛋糕的天使？慶幸曾經有這樣的天使出現過在我身邊，她就是戚黛黛。希望大家透過她這本書，找到開心、快樂的泉源！

蔡國威／六合彩王子

黛黛是我香港小姐選美前一屆的得獎者，認識她不知不覺已有10年的時間。還記得當年我參加選美的時候，她常來探望我們，為我們這班小師妹們打氣加油，分享她選美時的經驗。那時候，黛黛給我的印象就是位非常貼心、溫暖、善良的女生。

選美之後雖然我們見面的時間不多，但她的溫暖依然，我們每次碰面總有聊不盡的話題。曾經有好幾次她在電視裡看到我的演出，都會發來短訊問候和支持。這位美貌與智慧並重的香港小姐就是這樣默默地、溫柔地關愛著他人，發放著正能量。

去年，我應邀作嘉賓出席黛黛的電臺節目。她非常的體貼，在訪問的前一個星期就把訪問的內容、題目、時間和地點，以中英文形式整齊、清楚地通過電郵發送給我，在電郵最後還貼心地補充説：「訪問前先給妳參考一下，一切以保護受訪嘉賓隱私、讓嘉賓舒服為前提。」我當時想，她應該擔心我不太會看中文字，所以花了不少時間寫了中英文對照翻譯版本給我。無論何時，她都很照顧別人感受，在她身邊都會感到輕鬆、自在。

她做事專注，細心關懷他人，對待美食態度認真，更有自己獨特的見解。雖然我踏入餐飲業時間不長，但在經營管理時，都希望食物的質量和衛生能夠得到保證。而每當看到黛黛的美食作品時，都不禁讓我讚嘆，這不僅僅是視覺上的享受，同時也打開了觀者的味蕾，做到了色香味俱全。

我很榮幸能夠為黛黛的新書作序，在這裡我非常誠意地向大家推薦這樣一位懂吃、愛吃、用心的美食家。謹在此祝福黛黛和各位讀者甜蜜、幸福、順心、快樂。

朱慧敏／2004年香港小姐亞軍及著名演員

絲帶、蝴蝶結、手袋、cupcake都是代表黛黛。

以我非官方估計，黛黛應該是受了我的inspiration（又或是刺激）——連外行人都做了餅店，真正的行家怎麼能不出手？所以她終於將一本極具創意，很有風土特色的cupcake recipe獻給大家。當然我是「恨」了很久。現在的「獨樂樂，不如眾樂樂」的美麗分享只是小試牛刀，但絕不是初試啼聲，因為黛黛早已累積了不小實戰經驗。我知道，有一天，你必定會成為新一代的cupcake女神，擁有自己的boutique，跟大家分享精緻及具創意的甜點蛋糕。

Kelvin Tang／Founder of Treasure the Moment

一次江湖救急，黛黛成就了我人生中第一個親手製作的生日蛋糕，還給我家美人一次難忘的超級驚喜！雖然之前已經品嘗過這位才女的出品（實在是令人驚豔！），但從跟她一起做蛋糕的那天開始，我才真真正正感受到她對做蛋糕的那份熱忱和執著。這次她更以世界風情為藍本，自製特色cupcake攻陷讀者的心！就我本人來説，一口一個的cupcake，想來都相當美味……極之期待！撐！!!

姜皓文／藝人

認識黛黛超過10年，知道她一路以來都非常熱愛烹飪，而且充滿藝術感。去過她家的朋友，都會知道她家裡每一個角落都充滿色彩，還陳列了來自世界各地的有趣裝飾品。嘗過她做的蛋糕，無論是味道或者賣相，都很有心思和創意。對食物要求高又喜歡甜點和烹飪的我，看到這本書真的很興奮！我終於可以得到黛黛的祕笈！多謝她讓我的烹飪世界裡增加了很多色彩！

李施嬅／2003年香港小姐最上鏡小姐及著名演員

目錄

踏上Cupcake之旅，Are You Ready？

上路！把全世界的甜蜜裝進Cupcake！

香港・澳門 HONG KONG / MACAU

臺灣 TAIWAN

日本 JAPAN

新加坡・泰國 SINGAPORE / THAILAND

英國 UNITED KINGDOM

法國 FRANCE

義大利 ITALY

比利時 BELGIUM

美國 UNITED STATES

加拿大 CANADA

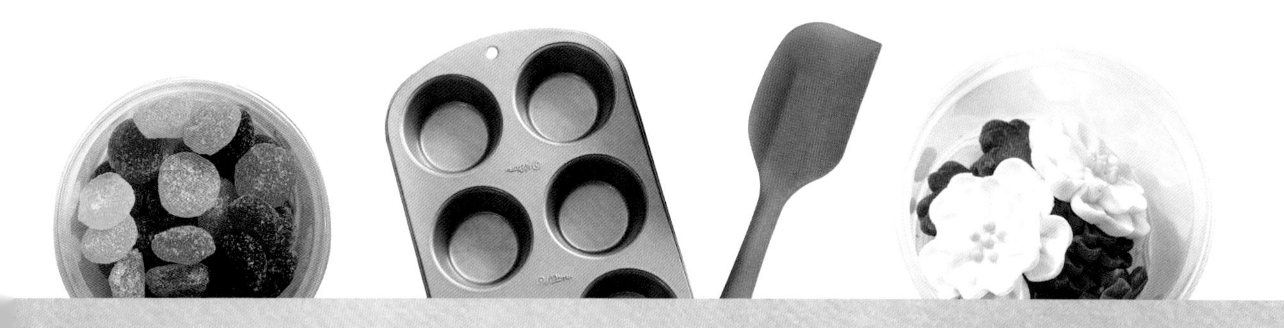

踏上Cupcake之旅
Are You Ready?

想在家自製Cupcake嗎？在動手做之前，

必須先了解常用工具、常用食材、單位及溫度換等基本知識，

對前製作業有了認識後，

再來學習基本蛋糕、糖霜與裝飾技巧，

相信你也能輕鬆做出好吃又好看的Cupcake！

★本書提供的食譜均為供製作6個蛋糕的份量。

All recipes in this book could be used to make 6 cupcakes.

常用工具 Basic Equipments

1.**量杯**：烹飪用的量杯，1杯等於250毫升。

2.**電子秤**：家用電子秤，量度單位為克／盎司。適用於量度烘焙用的乾、濕材料。

3.**粉篩**：用於篩撒麵粉、糖粉、可可粉等粉狀乾性烘焙材料。

4.**矽膠刮刀**：耐高溫、防沾，適用於一般需要用手攪拌的烘焙工序。

5.**不鏽鋼抹刀**：適合一般需要用手攪拌的烘焙工序，或用於塗抹奶油。

6.**攪拌碗**：材質有玻璃、不鏽鋼等，有不同尺寸可選擇，適合不同用途。

7.**電動攪拌器**：手提烘焙攪拌器，用於打蛋或打麵團，可根據不同工序調整速度快慢。

8.**馬芬杯蛋糕紙杯**：拋棄式防油蛋糕杯，可配合派對主題選擇不同顏色和形狀。

9.**馬芬蛋糕烤盤**：市場有多種不同材質可選，如傳統金屬烤盤、不沾烤盤或矽膠烤盤等。

10.**拋棄式擠花袋**：用於蛋糕裝飾，可用來擠奶油糖霜、奶油乳酪等。

11.**花嘴**：用於蛋糕裝飾，可使用不同花型花嘴做出不同形狀效果。

12.**剪刀**：用於剪裁擠花袋。

1.**Measuring cups:** Standard one cup=250ml in a glass measuring cup.
2.**Electric kitchen scale:** For measuring dry or wet ingredients for baking in unit –grams.
3.**Sifter / strainers:** For sifting dry ingredients, icing sugar, cocoa powder.
4.**Rubber spatula:** For folding or scraping mixture from bowls and stirring hot liquid.
5.**Metal spatula:** For normal mixing or cake frosting.
6.**Mixing bowls:** Stainless steel or glass, need a variety of sizes for hand mixing.
7.**Electric hand mixer:** Hand-held mixer with whisk attachment is most portable for mixing ingredients.
8.**Cupcake liners:** For baking and part of decorating tool. Can buy in variety of colors and shapes.
9.**Cupcake baking tray (muffin baking tray):** Comes in traditional stainless steel baking tray, non-stick aluminum baking tray, or silicone muffin baking tray.
10.**Disposable piping bags:** A tool to apply frosting or piping on cupcakes for decoration purpose.
11.**Piping tips:** Tools to create different designs for cake decorating.
12.**Scissors:** For cutting the piping bags.

蛋糕杯 Baking Cups

拋棄式防油紙杯：選用不同顏色、圖案和壓成不同形狀的烘焙用紙杯，可讓杯子蛋糕擁有百變外型，為派對增添色彩和歡樂氣氛。一般烘焙用紙杯都可於超市或烘焙專門店找到，適用於多數馬芬蛋糕烤模。

Paper Liners : A part of cake decorating process with the use of paper liners to match party & color themes. You can find it in local supermarkets or bakery stores and are available in different sizes, shapes, solid colors or special printed designs that are suitable for standard muffin pans or cupcakes pans.

耐高溫矽膠烤模：耐用、耐高溫，可重複多次使用的環保用具！矽膠是一種傳熱很快的材料，且有不沾的特性，很適合初學烘焙人士。

Silicone Liners : A good re-usable and environmental friendly choice for cake baking. They are durable, heat resistant and non-sticky when baking. Silicone is an excellent insulator that never gets hot while baking unlike traditional metal pans that will get very hot while in the oven.

常用食材 *Basic Ingredients*

自發粉：在麵粉中加入泡打粉，每110克加入1茶匙發粉便成自發粉。

Self-Raising Flour: plain flour sifted with baking powder in the ratio of 1 cup flour to 10g baking powder.

普通麵粉：低筋至中筋、無發酵粉的麵粉，蛋白質含量低，筋度、黏度較低，適用範圍廣。

Plain Flour: also known as all-purpose flour made from wheat.

巧克力：由可可豆提煉而成，分為白巧克力、牛奶巧克力和黑巧克力三類。可溶入烘焙蛋糕，或作巧克力味裝飾。

Chocolate: available in milk, white, dark chocolate. Ideal for decorating or melt the chocolate into the cake batter for a luxury rich cocoa flavor.

→白巧克力
↓黑巧克力

牛奶巧克力

蘇打粉（左）：蛋糕用膨脹劑的一種，多用於較帶酸性的蛋糕，與果汁、酸乳酪、巧克力等材料搭配，有中和和鬆軟的效果。

Baking Soda: also named bicarbonate of soda, a raising agent used in recipes with acidic ingredients such as lemon, chocolate, buttermilk, yogurt.

泡打粉（右）：蛋糕用發粉（膨脹劑），在烘焙加熱時，會膨脹產生鬆軟效果。

Baking Powder: a raising agent used mostly in cake baking and helps lighten the cake texture in baking.

可可粉：原味沒有添加糖粉，味道甘苦。可作蛋糕表面的裝飾。

Cocoa Powder: dried, unsweetened also known as cocoa.

黑糖：又名赤砂糖，帶有焦糖風味和不同深淺紅褐色。

Brown Sugar: comes in brown or dark brown sugar, a moist sugar with a rich full flavor & characteristic color.

白砂糖：白色、磨細的食用砂糖，可增加甜味，亦可在打發雞蛋時，起到穩定劑作用。

Caster Sugar: a soft finely granulated white table sugar.

香料粉：烘焙用香料粉，常用於提升餅乾的味道。常見的有肉桂粉、薑粉、豆蔻粉等。

Spice: grounded spices in powdered form, available in most supermarkets in a variety of different spices. Cinnamon, ginger, nutmeg is common used in festive cake or cookies recipes.

香精：常用香草精，用於增加香味。超市可買到不同果味的人工合成香精。

Extracts/Essence: non-alcoholic essences, which are available in different fruit flavours widely used in cake baking. Common extract used is vanilla extract made from vanilla beans.

轉化糖漿：可用玉米糖漿、楓糖漿、蜜糖等代替。

Golden Syrup: corn syrup, maple syrup or honey can be used as substitute.

糖粉：似粉狀的糖，呈白色粉末。市面購買的糖粉大多混入少許玉米澱粉作防潮之用。大多用於蛋糕裝飾。

Icing Sugar: confectioner sugar or powdered sugar, important base for buttercream or other frosting.

雞蛋：本書採用大號（L）美國雞蛋，以重量約50克作為食譜標準。

Eggs: use large egg average weighting 50g each (standard US size L egg).

奶油：市面分無鹽奶油和有鹽奶油兩大類，內含動物性脂肪。

Butter: salted or unsalted. Use unsalted butter for regular cake baking. It contains all animal fats.

奶油乳酪：用鮮奶油製作的新鮮軟質乾酪，具有濃郁的起司味和特殊酸味，通常用於製作起司類糕點。

Cream Cheese: a type of soft white smooth cheese, mild tasting with a high fat content.

白脫牛奶：又名酸奶或酪奶，低脂，質地似乳酪，味道微酸。內含乳酸，可讓蛋糕、麵包更鬆軟。

Buttermilk: a slightly sour thickened milk similar to yogurt. It's low in fat compared to fresh milk.

鮮奶油：較厚身的糖流質奶油，含脂肪最少35%或以上，打發後可作柔軟如雪糕般的擠花裝飾。

Whipping Cream: thickened cream with minimum fat content of 35%, good for making a soft unsweetened light whipped cream frosting for cupcakes.

食用色素：適合烘焙用之可食用人工色素，食品添加劑的一種。用於增強或改良食物顏色外觀。

Food Colourings: edible substance used to give colors to food, available in liquid, powdered or gel form.

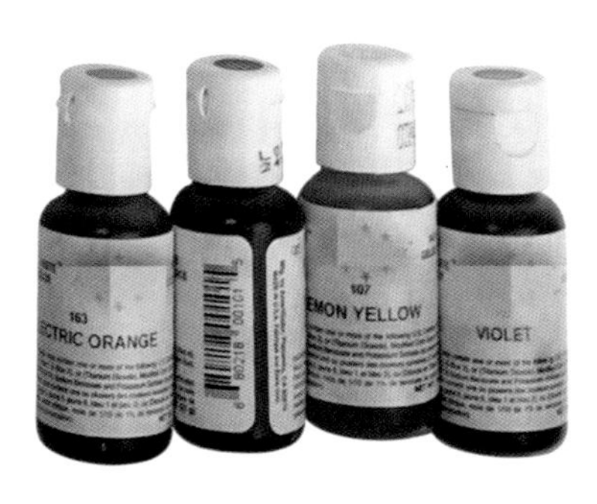

植物油：由不同植物提煉，市面上常見的植物油有蔬菜油、玉米油、橄欖油等。

Vegetable Oil: or salad oil, sourced from plant, it helps creates lighter sponge texture in cake baking.

牛奶：一般飲用鮮奶分為全脂、低脂和脫脂三種。市面上也有售添加不同果汁的果味奶。

Milk: use full cream fresh milk or skimmed milk, it comes in different fruity flavors.

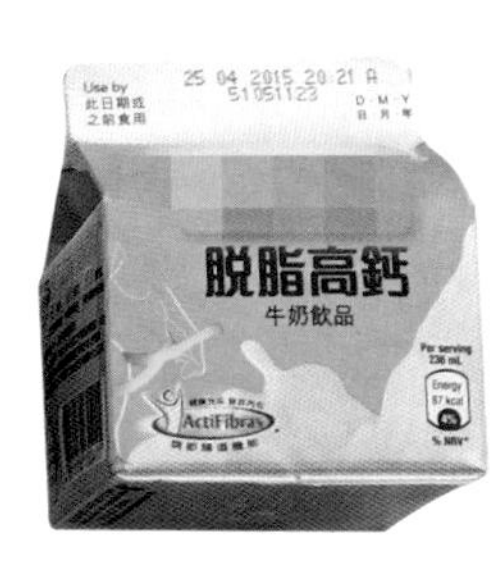

單位及溫度換算

Conversion Chart & Oven Temperature

換算表 CONVERSION CHART

小提醒：將所有材料準確地用量杯或電子秤量好備用。材料如需用量杯、茶匙、湯匙等量度，注意先用湯匙底輕手抹平多餘份量。

Useful hints: all cup & spoon measurements need to be measured in level. The accurate way to measure all ingredients is to weigh them with a metric marked glass cup or electric kitchen scale.

乾材料 DRY ingredients			濕材料 LIQUID ingredients		
公制計量 Metric	英制計量 Imperial		公制計量 Metric	英制計量 Imperial	
15克	1/2盎司		30毫升	1盎司	
30克	1盎司		60毫升	2盎司	
60克	2盎司		100毫升	3盎司	
90克	3盎司		125毫升	4盎司	
125克	4盎司	1/4磅	150毫升	5盎司	
250克	8盎司	1/2磅	250毫升	8盎司	
375克	12盎司	3/4磅	300毫升	10盎司	
500克	16盎司	1磅	500毫升	16盎司	1磅
750克	24盎司	1又1/2磅	600毫升	20盎司	1品脫
1公斤	32盎司	2磅	1000毫升	1又3/4品脫	1公升

克（g）；公斤（kg）；盎司（oz）；磅（lb）；毫升（ml）；品脫（pint）；公升（litre）

烤箱溫度 OVEN TEMPERATURE

小提醒：注意不同牌子的家用烤箱溫度計算有分別，請先參考烤箱說明書。較小型家用烤箱可調較溫度低10度和烘焙時間縮短5～7分鐘左右。

Useful hints: Guide to conventional ovens (for fan-forced oven, check manufacturer's menu). For small home use oven, you can adjust the baking temperature to 10 degrees lower and shorten the baking time by 5-7 minutes.

攝氏 C-Celsius	華氏 F-Fahrenheit	烤箱溫度指數對照 Gas mark
120	250	1/2
150	275～300	1～2
160	325	3
180	350～375	4～5
200	400	6
220	425～450	7～8

基本蛋糕製作方法 Basic Cupcake Recipes

基本香草奶油蛋糕 Basic Vanilla Cupcakes

材料

- □無鹽奶油 45克
- □白砂糖 70克
- □雞蛋 1顆
- □自發粉 90克
- □脫脂牛奶 40毫升
- □植物油 20克
- □香草精 1/2茶匙

步驟

1 **準備：**將烤箱預熱至170℃。奶油在室溫中回軟。

2 **打發奶油：**加入白砂糖，和奶油一起打發至顏色變淺，呈軟滑狀。

3 **加入雞蛋：**加入打散的蛋液，均勻攪拌，直到和奶油混合。

4 **加入過篩的粉類：**將自發粉過篩，然後分3次加入。用矽膠刮刀以切拌的方式快速拌勻。

5 **加入其他材料：**加入脫脂牛奶，攪拌均勻，再拌入植物油和香草精。

6 **入烤箱烘焙：**在馬芬杯裡放入紙模，倒入麵糊，約2/3滿。用170℃烤20～30分鐘或烤至蛋糕表面呈金黃色。

小技巧：如何判斷蛋糕是否熟透了？

在蛋糕中間插入牙籤，取出後如果牙籤有些濕濕的，代表蛋糕還未完成。相反，如果牙籤是乾身且有蛋糕碎的話，表明蛋糕已經熟透了。

INGREDIENTS

45g unsalted butter (at room temperature)
70g caster sugar
1 large egg
90g self-raising flour, sifted
40ml skimmed milk
20g vegetable oil
1/2 teaspoon vanilla essence

STEPS

1. Preheat the oven to 170 degrees Celsius.
2. Beat the butter and sugar until light yellow color and smooth.
3. Add in the egg and mix well.
4. Sift the self-raising flour. Add in the flour little by little to the batter.
5. Add in the milk little by little and then add in vegetable oil and the vanilla essence.
6. Pour into cupcake cups, 2/3 full. Bake for 20-30 minutes or until golden brown.

TIPS

How to know if the cupcakes are ready?

Insert a toothpick in the middle of the cupcake and pull it out. If there are some watery fluid cupcake batter stuck on the toothpick, it means it is not ready. However, if the toothpick is dry with some cupcake pieces on it, then it is ready!

基本巧克力蛋糕 Basic Chocolate Cupcakes

材料

- 無鹽奶油 55克
- 黑糖 70克
- 蘇打粉 1/2茶匙
- 雞蛋 1顆
- 泡打粉 1/8茶匙
- 麵粉 90克
- 白脫牛奶 60毫升
- 黑巧克力 30克
- 可可粉 1湯匙（先溶於1湯匙熱水內）

步驟

1 **準備**：將烤箱預熱至170℃。奶油在室溫中回軟。

2 **打發奶油**：用電動攪拌器打發奶油，加入黑糖，繼續打至顏色變淺，呈軟滑狀。

3 **加入雞蛋**：加入打散的蛋液，均勻攪拌，直到和奶油混合。

4 **篩入粉類**：將麵粉、蘇打粉和泡打粉混合過篩，然後分3次加入。

5 **加入白脫牛奶、可可粉醬**：依次加入白脫牛奶和可可粉醬。

6 **加入巧克力**：拌入融化的黑巧克力。

7 **入烤箱烘焙**：在馬芬杯裡放入紙模，倒入麵糊，約2/3滿。用170℃烤20～30分鐘或烤至蛋糕表面呈金黃色。

INGREDIENTS

55g unsalted butter (at room temperature)
70g brown sugar
1/2 teaspoon baking soda
1/8 teaspoon baking powder
1 large egg
90g all purpose flour (sifted)
60ml buttermilk
1 tbsp cocoa powder (melted in 1 tbsp of hot water)
30g dark chocolate chips (melted and folded in batter)

STEPS

1.Preheat the oven to 170 degrees Celsius.
2.Beat the butter and sugar until light yellow in color and smooth.
3.Add in the egg and mix well.
4.Sift the flour, baking soda and baking powder together. Add in the flour little by little to the batter.
5.Add in the buttermilk and cocoa paste little by little.
6.Stir in the melted chocolate.
7.Pour into cupcake cups, 2/3 full. Bake for 20-30 minutes or until golden brown.

基本糖霜製作方法 Basic Icing Recipes

基本奶油糖霜 Basic Buttercream

材料

- 無鹽奶油 100克
- 糖粉 200克
- 香草精 1茶匙
- 冷水或脱脂牛奶 2～3茶匙

步驟

1 **打發奶油和糖粉**：奶油在室溫中回軟，用攪拌器打至軟滑，慢慢篩入糖粉，繼續攪拌。

2 **加入液體**：慢慢加入冷水或牛奶，再加入香草精。

3 **繼續打發**：繼續打發至鬆軟、無粉粒狀態。

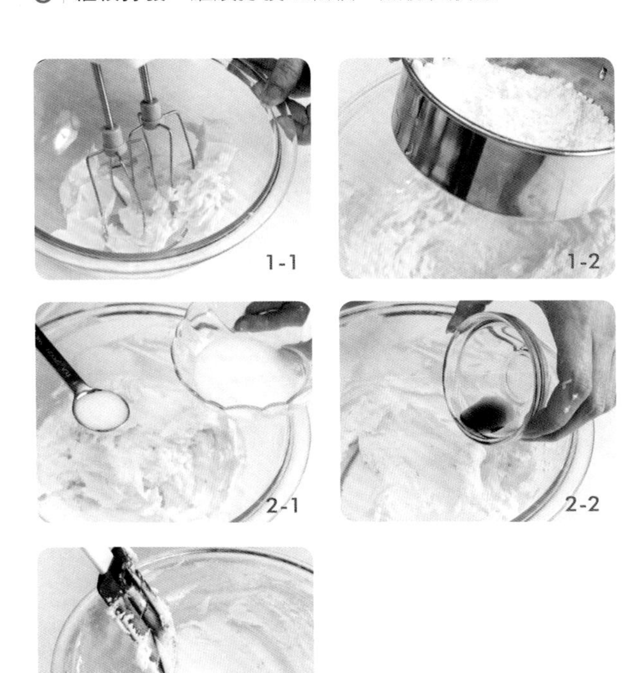

INGREDIENTS

100g unsalted butter (at room temperature)
200g icing sugar (sifted)
2-3 teaspoons of skimmed milk or cold water
1 teaspoon vanilla essence

STEPS

1.Beat the butter at medium high speed until soft and creamy. Add in the sifted i cing sugar little by little, beat at high speed until all blend in.
2.Add in the milk or cold water one tablespoon at a time.
3.Beat at high speed until icing is smooth and fluffy.

TIPS

1.Store in air tight container inside fridge for 4-5 days.
2.Bring to room temp before frosting on cakes.

小技巧

1.將奶油糖霜放入密封容器，可於冰箱冷藏室儲存4～5天。
2.擠花前，應先將奶油糖霜放至室溫，才開始擠上蛋糕表面。

蛋白霜 Meringue Frosting

此食譜蛋白霜質感輕盈，入口感軟滑如雪糕般！

材料

糖　漿：□白砂糖 165克　□玉米糖漿 15克（1湯匙）　□冷水 30克

蛋白霜：□白砂糖 15克（6茶匙）　□蛋白 99克（大號雞蛋 3顆）

步驟

1. **煮沸糖水：**在小鍋中放入水、玉米糖漿和白砂糖，煮沸至110℃。
2. **打發蛋白：**將蛋白放入另一個打蛋盆中，用攪拌器以中速打發出泡沫，分多次慢慢加入砂糖，繼續用中高速打發蛋白到硬性發泡。
3. **混合：**將煮沸糖水慢慢倒入蛋白霜內，繼續用高速打發蛋白霜至光亮，鬆軟順滑，待蛋白霜降回室溫，才可放入擠花袋開始擠花步驟。

1-1

1-2

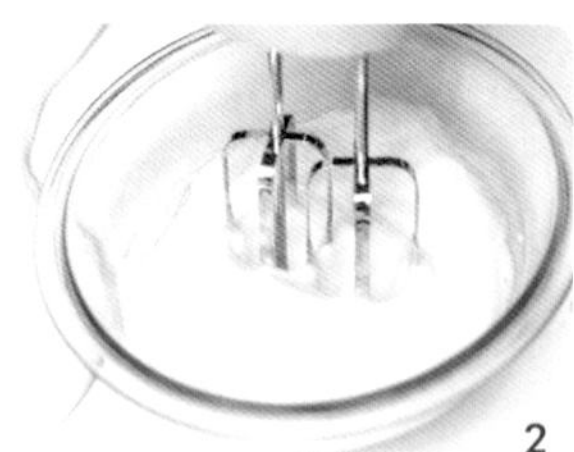
2

3

INGREDIENTS

Syrup:
165g granulated sugar
15g (1 tbsp) corn syrup
30g water
Meringue:
99g (3 large sz) egg white
15g (6tsp) granulated sugar

STEPS

1.Bring icing sugar, corn syrup, water in a saucepan to broil at 240F (110 degrees Celsius).
2.In another bowl, beat egg whites & white sugar to firm peaks.
3.Pour in hot syrup in a slow stream down from side of bowl into egg white, and beat on high speed until meringue looks fluffy and glossy, cool down before frosting on cupcakes.

基本糖霜製作方法 Basic Icing Recipes

法式蛋白奶油霜
French Meringue Buttgercream Frosting

材料

- 白砂糖 45克
- 無鹽奶油 110克
- 蛋白 40克（超大雞蛋 1顆）

步驟

1 **打發蛋白：**將蛋白放入打蛋盆，用攪拌器中速打發蛋白成泡沫狀。之後慢慢加入砂糖，繼續高速打至硬性發泡。

2 **加入奶油：**加入室溫奶油，要分多次拌勻入蛋白霜內（30克，約2湯匙量）。

3 **繼續打發至濃稠：**攪拌器繼續高速打發至奶油霜變成濃稠順滑。

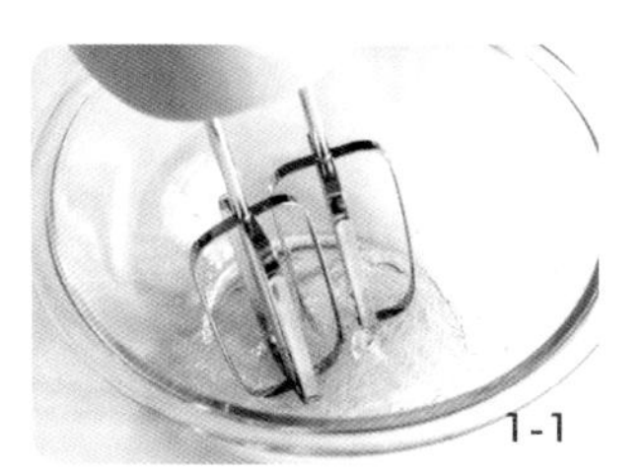
1-1

1-2

2

3

INGREDIENTS
40g (1xLsz) egg white
45g sugar
110g unsalted butter
(at room temperature)

STEPS
1.In a bowl, beat egg whites to foamy. Add in sugar and beat to stiff peaks.
2.Add in butter (soft room temp) 2 tablespoons each time into the meringue.
3.Continue to beat at high speed until buttercream starts to thicken and becomes smooth.

TIPS
When you first add in the butter, meringue will break down and curd. Keep beating in butter at high speed and buttercream will form after 10-15 minutes time.

小技巧
當加入軟奶油粒時，蛋白霜與奶油會形成分離狀態，這是正常的。繼續打發約5～10分鐘後，蛋白奶油會自然混勻。

奶油乳酪糖霜 Cream Cheese Frosting

材 料

□奶油乳酪 100克 □糖粉 100克 □香草精 1/4茶匙（可選擇不加入）

步 驟

1 | **打發奶油乳酪**：將室溫奶油乳酪放入打蛋盆中，用電動攪拌器以中速打至鬆軟。

2 | **加入糖粉**：將糖粉過篩，分6~7次慢慢加入。

3 | **繼續打發**：用電動攪拌器繼續打發至鬆軟順滑。

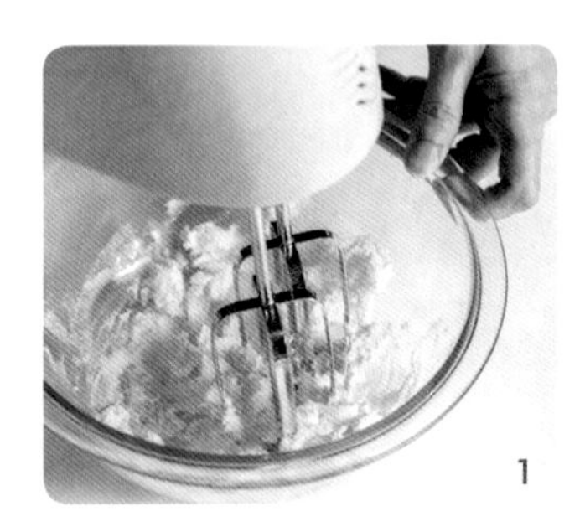

INGREDIENTS

100g cream cheese (at room temperature)
100g icing sugar (sifted)
1/4 teaspoon vanilla extract (optional)

STEPS

1.Beat the cream cheese until soft and creamy.
2.Add in the sifted icing sugar little by little.
3.Beat at high speed until icing is smooth and fluffy.

基本糖霜製作方法 Basic Icing Recipes

巧克力醬 Chocolate Ganache

材料

□鮮奶油 100克　□60%黑巧克力 200克

★巧克力：鮮奶油＝2：1（若用於淋面可調整至1：1或1：1.5）

步驟

1 **加熱鮮奶油：**將鮮奶油倒入小碗內，放入微波爐加熱2～3分鐘至邊緣冒泡微沸狀態。

2 **融化巧克力：**巧克力置於另一碗中，倒入熱的鮮奶油，靜待2～3分鐘至巧克力開始融化。

3 **混合：**用矽膠刮刀輕輕將巧克力和奶油攪拌至完全均勻，呈光亮軟滑狀。

4 **完成：**完成後放入擠花袋，或可用抹刀將巧克力醬抹平於蛋糕表面。

INGREDIENTS

100g whipping cream
200g 60% dark chocolate
* Ratio – chocolate 2 : heavy whipping cream 1 (adjust to 1:1 / 1:1.5 if you need a more softer creamy consistency)

STEPS

1. Microwave whipping cream for 2-3 minutes until hot but not boiling.
2. Pour hot cream into chocolate buttons until it melts.
3. Stir with spatula until it all blend in to form a smooth paste.
4. Place into piping bag or pour onto the cupcakes.

TIPS

1. Use immediately and pour onto the cupcakes if you want a liquid consistency.
2. However, if you need a firmer consistency for piping or frosting, cover the chocolate with plastic wrap and chill it in the fridge for 30 minutes. (Check every 10 minutes to see if the Ganache has already reached the right consistency for use or else it will be become too hard.)

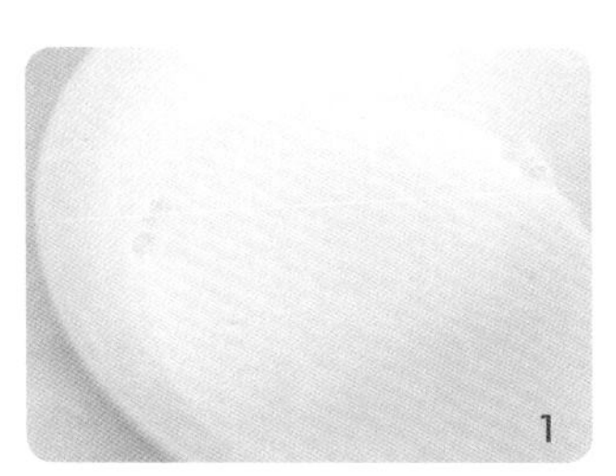

1

2-1

2-2

3

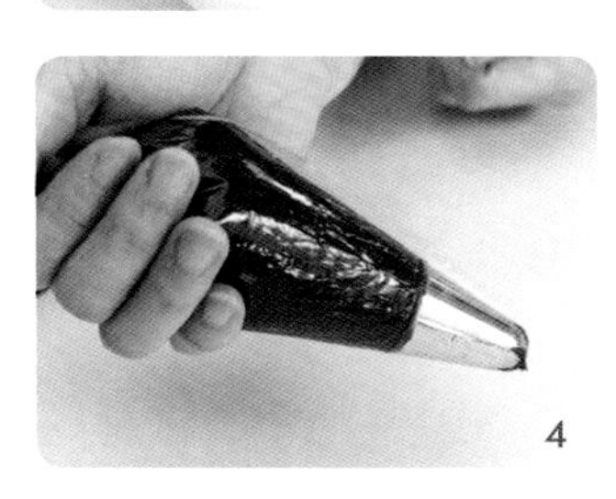

4

小技巧

1. 如果需要鏡面淋醬效果，可立即淋上蛋糕表面待涼。
2. 如需要做擠花塑形，將巧克力醬蓋一張保鮮膜放入冰箱約半小時（約10分鐘時，可查看巧克力醬是否達到理想軟硬度，若冷藏太久會變硬，不適合擠花）。

預備擠花袋 *Fill a Piping Bag*

工具

□ 拋棄式擠花袋 □ 花嘴 □ 抹刀 □ 高腳玻璃杯 □ 剪刀
□ 橡皮筋

步驟

1. **準備擠花袋**：預備1個拋棄式擠花袋，在袋尾尖處剪1個約1～1.5吋的開口（約花嘴長度一半）。
2. **放入花嘴**：將花嘴放入擠花袋內。花嘴較窄的一端要突出於擠花袋口外。
3. **套入玻璃杯**：將擠花袋套入1個高腳玻璃杯，擠花袋口向下包住玻璃杯口。
4. **放入奶油糖霜**：以抹刀將預備好的奶油糖霜放入袋內，用手將奶油糖霜推至袋尖，並推出袋內空氣。
5. **密封**：用手將擠花袋口按實，以橡皮筋將擠花袋口密封。

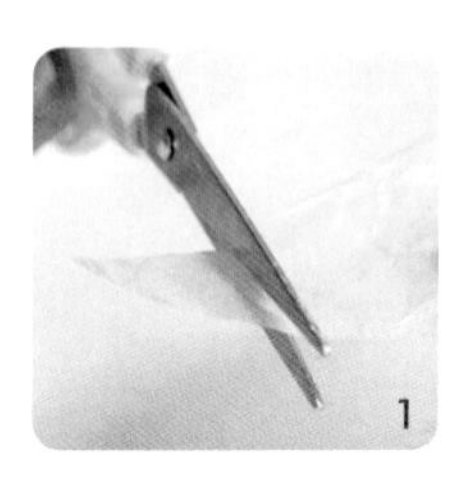
1

2

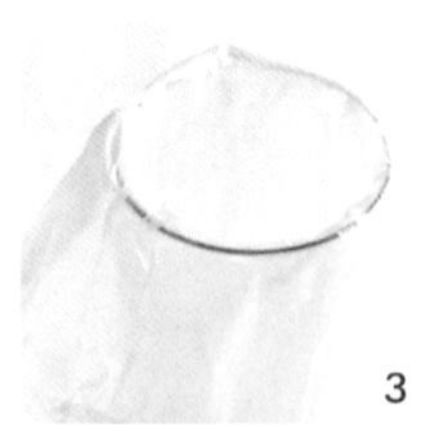
3

4-1

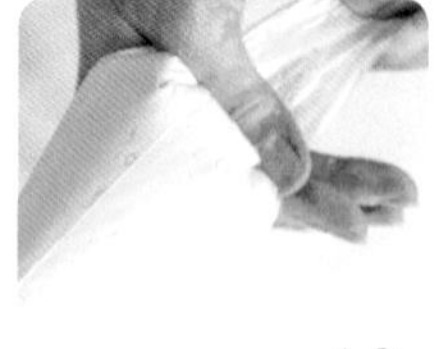
4-2

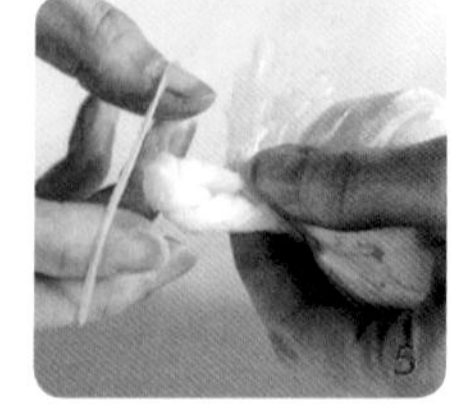
5

TOOLS

disposable plastic piping bag (or you can use zip-lock bag at home)
piping tip
spatula
a tall glass
scissors
rubber band

STEPS

1. Get a disposable piping bag, use scissors to cut out the narrow tip off with 1 to 1.5 inches (depending on the tip size).
2. Select a piping tip to be used and drop it into your piping bag. Fit the tip to the narrow end of the piping bag.
3. Put the bag into a tall glass and fold the outside of the piping bag over the lip.
4. Use a spatula and fill in frosting into piping bag. Use your hand to try scraping down the icing to the bottom of the tip to get rid of all the air bubbles.
5. Lift the bag out of the glass, use your hand to twist the top of the piping bag and tie it up with a rubber band. This can avoid icing spread out during piping.

基本裝飾技巧 Basic Decoration Tips

調色技巧 Make Color Icing

材料

□ 奶油糖霜　□ 調色杯　□ 食用色素　□ 牙籤　□ 抹刀

步驟

1 **打發奶油**：先打發好適量白色奶油糖霜備用。

2 **混入色素**：用牙籤挑出少量色素，再混入奶油糖霜中。

3 **拌勻**：用抹刀以畫圈式將顏色拌勻於糖霜內。

4 **重複步驟**：如要將顏色再調深，可用1支新牙籤挑出顏色重複步驟2、3。

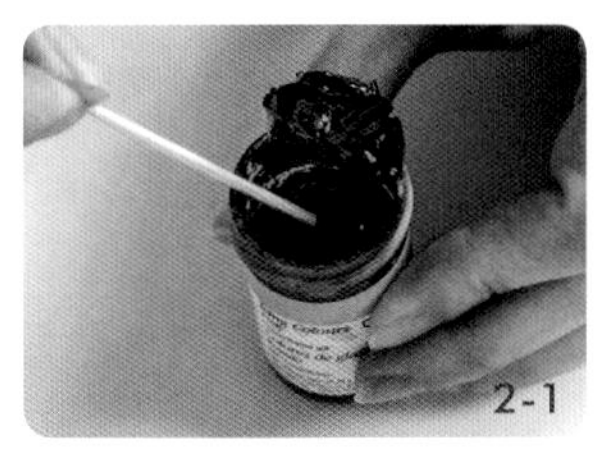
2-1

2-2

4-2

3

4-1

INGREDIENTS

buttercream icing
mixing cups
edible food color
toothpicks
spatula

STEPS

1.Start with basic original white buttercream icing.
2.Insert toothpick into the food coloring bottle to pick up a small amount and swirl into icing.
3.Use a spatula to blend in color to icing.
4.Use a new clean toothpick to add more color if the color is not enough.

花嘴使用技巧 Piping Tips

常見花嘴

1.多瓣花齒花嘴：可擠出較動感的波浪紋線條效果。

2.開縫星齒花嘴：可擠出較細緻的扭繩紋或立體多層冰淇淋效果。

3.圓口花嘴：可擠出蜂巢圈效果或用特小圓口花嘴作寫字劃線裝飾。

4.開縫法式花嘴：可擠出蓬鬆狀的花紋。

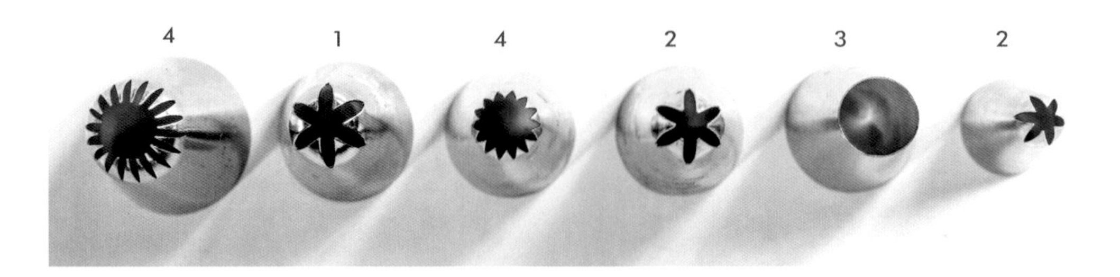

HELPFUL HINTS TO CHOOSE A TIP

1.Use a Closed Star Piping Tip: If you want a shorter lower frosting or more"ruffly"look swirls.
2.Use a Open Star Piping Tip: If you want a taller, ice cream frosting or a thicker rope look.
3.Use a Round Piping Tip: If you need to draw a line or shapes on cupcake design.
4.Use a French Star Piping Tip: If you want to pipe fluffy curls, or shell patterns.

基本裝飾技巧 Basic Decoration Tips

漩渦狀裝飾 The Perfect Swirl

工具

多瓣花齒花嘴（使用不同大小的花嘴及花齒可做出不同效果）

步驟

1 **擠花袋垂直於蛋糕表面：**注意擠花袋位置要垂直於蛋糕表面成90度角。

2 **控制擠壓力度：**拇指、食指按緊擠花袋頂部，其餘手指平均按緊擠花袋，以控制擠出的糖霜量。

3 **擠出第1層：**將手按緊擠花袋頂部，於蛋糕中心位置擠出適量糖霜，再轉1小圈。繼續由內圈擴展至蛋糕外圈形成第1層。

4 **重複：**用相同擠法再疊上第2、第3層。當到達最頂層，放鬆手指，停止按壓擠花袋及向上抽離。

TOOL

A medium size Closed Star Piping Tip (or use different tips using same technique)

STEPS

1. Hold piping bag in vertical up right position to your cupcake.
2. Apply even pressure to your frosting bag.
3. Start piping out from the center of cupcake. Pipe a large circle from centre out and around the outer-edge of cupcake.
4. Continue piping smaller circles on top of one another. Release pressure (stopsqueezing the bag) then lift up.

星形花裝飾 Star Drop Flower

工具

小型或中型開縫法式花嘴

步驟

1 **擠花袋垂直於蛋糕表面：**注意擠花袋位置要垂直於蛋糕表面成90度角。

2 **控制擠壓力度：**拇指、食指按緊擠花袋頂部，其餘手指平均按緊擠花袋，以控制擠出的糖霜量。

3 **擠出星形花：**手指平均用力，按緊擠花袋頂部，擠出適量糖霜，於蛋糕沿外圍完成1圈星形花。當完成每1粒星形花，放鬆手指，停止按壓擠花袋及向上抽離。

4 **重複：**重複以上步驟，直至星形花平均遮蓋整個蛋糕表面。

1

2

3-1

3-2

4-1

4-2

TOOL

A small to medium French Star Piping Tip

STEPS

1. Hold the piping bag in upright position, then Apply pressure and start at the outer edge of the cupcake.
2. Apply even pressure to your frosting bag.
3. Let the icing come out quickly, move tip upwards and stop the pressure.
4. Repeat the process for the entire cupcake and finish the last one for the centre.

基本裝飾技巧 Basic Decoration Tips

蓬鬆球裝飾 Fluffy Curl

工具

中型或大型開縫法式花嘴

步驟

1 **擠花袋垂直於蛋糕表面：**注意擠花袋位置要垂直於蛋糕表面成90度角。

2 **控制擠壓力度：**拇指、食指按緊擠花袋頂部，其餘手指平均按緊擠花袋，以控制擠出的糖霜量。

3 **擠出蓬鬆球：**手指平均用力，按緊擠花袋頂部，於蛋糕中心位置擠出適量糖霜。繼續輕力按壓擠花袋，並將手慢慢提升擠出糖霜，讓中心疊起形成1個微捲小球。

4 **完成：**當擠至需要高度，放鬆手指，停止按壓擠花袋及向上抽離。

1

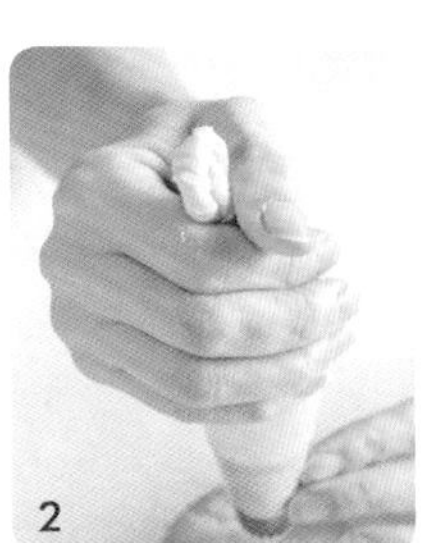

2

3

4-1

4-2

4-3

TOOL

A medium or big French Star Piping Tip

STEPS

1. Hold the piping bag vertically and apply the pressure from the center of the cupcake.
2. Apply even pressure to your frosting bag.
3. Apply more pressure, let the icing come out and raise the tip slightly in a slow motion.
4. Stop the pressure when it reaches the tall end and lift the tip straight up.

玫瑰花裝飾 *Rose*

工具

中型或大型開縫星齒花嘴

步驟

1. **擠花袋垂直於蛋糕表面：**注意擠花袋位置要垂直於蛋糕表面成90度角。
2. **控制擠壓力度：**拇指、食指按緊擠花袋頂部，其餘手指平均按緊擠花袋，以控制擠出的糖霜量。
3. **擠出玫瑰花：**手指平均用力，按緊擠花袋頂部，擠出適量糖霜於蛋糕中心位置轉1小圈。繼續擠出糖霜，沿內圈向外繞出1～2圈，直到離蛋糕紙杯邊約0.5～0.8公分。
4. **完成：**當完成最後外圈，放鬆手指，停止按壓擠花袋及向側面抽離。

TOOL

A medium or large Open Star Piping Tip

STEPS

1. Hold the piping bag in a vertical up right position.
2. Apply a slight pressure and make a dollop in center.
3. Without releasing the pressure, slowly move tip out from center and start make circle around center dollop. Complete with one to two circles (depending on the size of the star piping tip you use)around the center.
4. Finish by sliding the tip down to the outer-edge of cupcake and release the pressure and pull the tip off.

基本裝飾技巧 Basic Decoration Tips

小花群裝飾 *Blossoms*

工具

小型或中型開縫星齒花嘴

步驟

1 **擠花袋垂直於蛋糕表面：**注意擠花袋位置要垂直於蛋糕表面成90度角。

2 **控制擠壓力度：**拇指、食指按緊擠花袋頂部，其餘手指平均按緊擠花袋，以控制擠出的糖霜量。

3 **擠出小花：**手指平均用力，按緊擠花袋頂部，擠出適量糖霜，沿蛋糕外圍成1粒粒花形效果。每完成1粒小花，放鬆手指，停止按壓擠花袋及向上抽離。

4 **重複：**重複以上步驟至小花群均勻遮蓋整個蛋糕表面。當完成最外圍，放鬆手指，停止按壓擠花袋及向側面抽離。

1

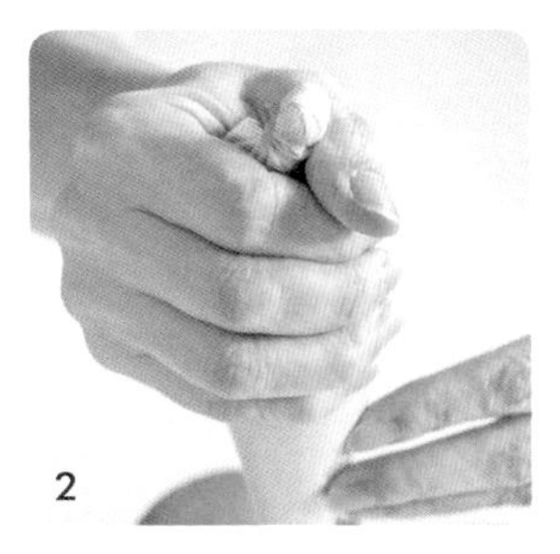
2

3

4

TOOL

A small or medium Open Star Piping Tip

STEPS

1. Hold the piping bag in an up-right position.
2. Start piping out the blossom from the outer circle of the cupcake.
3. Apply pressure to let the icing come out to form a blossom, then quickly lift the tip up and release pressure on the bag.
4. Repeat that same motion around the entire cupcake. Finish the last blossom in the centre.

盤繞線圈裝飾 *Coil*

工 具

大圓口花嘴

步 驟

1. **擠花袋垂直於蛋糕表面：**注意擠花袋位置要垂直於蛋糕表面成90度角。
2. **擠出線圈：**手指平均用力，按緊擠花袋頂部，擠出適量糖霜，螺旋圍繞中心點，並繼續圍繞著蛋糕的邊緣，擠1圈。
3. **擠出第2圈：**回到起點，繼續擠第2圈。輕輕重疊於第1圈之上，做成類似線圈的樣子。
4. **重複：**重複同樣的步驟，最後放鬆手指，向上拉起，形成1個尖嘴。

1

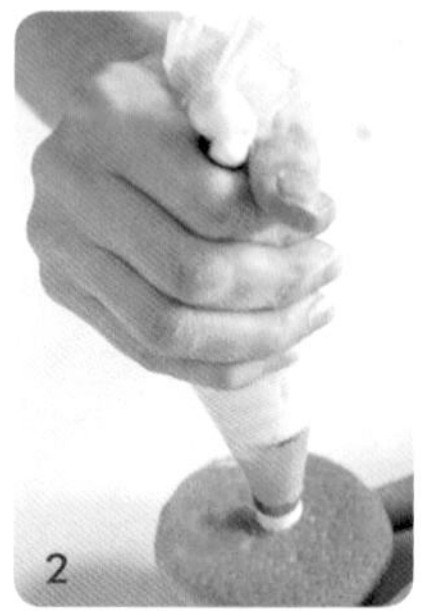
2

3-1

3-2

4-1

4-2

TOOL

A large Round Piping Tip

STEPS

1. Start at the center of your cupcake. Hold the piping bag in vertical up right position.
2. Apply even pressure. Pipe a spiral around center point and continue to make a circle around the border of the cupcake.
3. When you get to the starting point, continue piping the second circle slightly overlapping the first circle to create a coil look.
4. Repeat the same process, continue to swirl the top, release pressure and pull tip up to form a soft peak.

基本裝飾技巧 Basic Decoration Tips

簡易蛋糕裝飾 Simple decoration

利用不同形狀、顏色的糖果，
可將普通的杯子蛋糕裝點出不同的風格。

It is a simple, unique way to use candies, chocolates, sugar beads in all shapes and sizes as decorative toppers to create your own frosted cupcakes with an individual design.

派對紙製裝飾
Party paper toppers

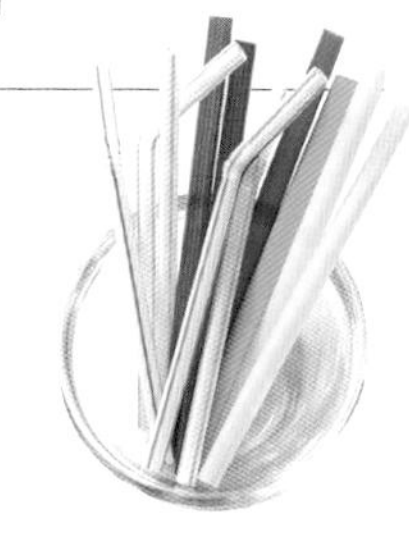

彩色吸管
Drinking straws

皇室糖霜小花
Sugar flowers

彩色巧克力豆
Chocolate beans

棉花糖
Marshmallows

食用裝飾銀珠糖
Silver sugar beads

黑巧克力薄片
Dark chocolate flakes

紅色裝飾彩砂糖
Red color sanded sugar

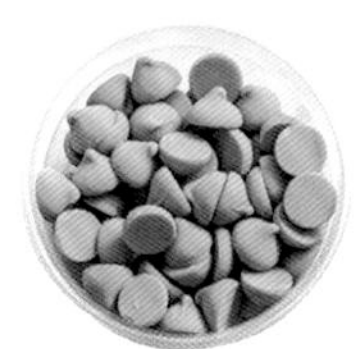

薄荷巧克力粒
Mint flavor chocolate chips

彩色圓形小糖珠
Round sugar candies

綠色裝飾彩砂糖
Green color sanded sugar

彩色迷你糖珠
Rainbow sugar beads

白巧克力粒
White chocolate buttons

迷你彩色巧克力豆
Mini chocolate beans

草莓味花瓣巧克力片
Strawberry flavored white chocolate curls

黑巧克力粒
Dark chocolate chips

迷你黑巧克力粒
Mini dark chocolate chips

天然免烤碎杏仁粒
Dried chopped almond

花瓣白巧克力片
White chocolate curls

糖衣軟糖
Sugar coated jelly beans

黃色裝飾彩砂糖
Yellow color sanded sugar

皇室糖霜紅蘿蔔裝飾
Sugar carrots toppers

上路！把全世界的甜蜜裝進Cupcake

旅行之中你最想記錄的是什麼？

香港、澳門、臺灣、日本、新加坡、泰國、

英國、法國、義大利、比利時、美國、加拿大！

從12個美食之都汲取靈感，

最能體現各地特色的cupcake上場囉！

香港・澳門
HONG KONG MACAU

香港是我第2個家，和我一樣都擁有中西文化兩種背景。
香港與澳門過去都曾是殖民地，
在文化、飲食、建築等等，都會受到曾統治的國家影響。
比如香港的街頭小吃，既有魚蛋這種中國傳統美食，
也有蛋塔、雞蛋仔這類經過了本地化的西式食物；
澳門的特產有保留中式傳統的杏仁餅、蛋捲、鳳凰卷，
也有具有葡萄牙特色的葡式蛋塔、木糠布丁等等。
許多人喜歡去澳門的原因，
除了光顧賭場，還有享受別具風格的當地美食。

芝麻蛋糕

芝麻糊是中國傳統甜點之一，香港很多甜點店都可以吃到。但是，有時總會覺得芝麻糊比較甜膩，如果吃1整碗的話，很容易感到飽足。這款黑芝麻cupcake的內餡和糖霜都加了芝麻糊，充滿濃香的芝麻滋味，又避免了飽腹感，適合喜愛芝麻口味的你！

材料

蛋糕

- 無鹽奶油 55克
- 白砂糖 50克
- 雞蛋 1顆
- 自發粉 100克
- 芝麻粉 1.5湯匙
- 脱脂牛奶 40毫升
- 植物油 25克
- 烤黑芝麻粒 10克
- 黑色食用色素 1滴

芝麻奶油糖霜

- 無鹽奶油 100克
- 脱脂牛奶 2～3茶匙
- 糖粉 200克
- 芝麻粉 2湯匙
- 黑色食用色素 1滴

裝飾

- 烤黑芝麻粒 適量
- 擠花袋 1個
- 花瓣花嘴 1個

製作蛋糕

1 **準備：**將烤箱預熱至170℃。奶油置於室溫回軟。

2 **打發奶油：**加入白砂糖，和奶油一起打發，直到顏色變淺，呈軟滑狀。

3 **加入雞蛋：**加入打散的蛋液，均勻攪拌，直到和奶油混合。

4 **篩入粉類：**將自發粉過篩，然後分3次加入。用矽膠刮刀以切拌的方式快速拌勻。

5 **加入其他材料：**依次加入芝麻粉、牛奶、植物油、烤黑芝麻和黑色食用色素。

6 **入烤箱烘焙：**將麵糊慢慢倒入蛋糕杯，約2/3滿。用170℃烤20～30分鐘或烤至蛋糕表面呈金黃色。

7 **冷卻：**完成後，轉移到散熱架上。等蛋糕完全冷卻後才裝飾。

製作芝麻奶油糖霜

1 **打發奶油**：先將室溫奶油用電動攪拌器打至軟綿，分5～6次加入糖粉。

2 **加入其他材料**：加入脫脂牛奶、芝麻粉，然後拌勻。加入黑色食用色素。

3 **攪拌均勻**：將所有材料打發至鬆軟、無粉粒狀態。

1

2-1

2-2

3

裝飾步驟

1-1 1-2 2

1 **使用花瓣花嘴**：拿1個擠花袋，放入花瓣花嘴。將芝麻奶油糖霜放入擠花袋，往下推實，紮緊袋口，在蛋糕表面擠上花紋。開始擠花時，花嘴要和蛋糕成45度角，輕輕擠出糖霜，手微向前後推動呈小波浪動作。先完成外圈圍邊再繼續完成內層。

2 **撒上芝麻粒**：最後撒上烤黑芝麻粒作裝飾。

Sesame Cupcakes

INGREDIENTS

For Cupcake

55g unsalted butter
(at room temperature)
50g caster sugar
1 large egg
100g self-raising flour
1.5 tablespoon instant sesame dessert powder
40ml skimmed milk
25g vegetable oil
10g toasted black sesame
1 drop of black food coloring

For Sesame Buttercream

100g unsalted butter
(at room temperature)
200g icing sugar
2-3 teaspoons skimmed milk
2 tablespoons instant sesame dessert powder
1 drop of black food coloring

For Decoration

Petal Piping Tip – ruffle flower
1 piping bag
toasted black sesame sprinkle on top

CUPCAKE STEPS

1. Preheat the oven to 170 degrees Celsius.
2. Beat the butter and sugar until smooth.
3. Add in the egg and mix well.
4. Sift the self-raising flour. Add in the flour and sesame dessert powder little by little to the batter.
5. Add in the milk and vegetable oil little by little. Stir in the toasted black sesame. Add in a drop of black food coloring.
6. Pour into cupcake cups, 2/3 full. Bake for 20-30 minutes or until golden brown.
7. Transfer to cooling rack. Cool down completely before frosting.

SESAME BUTTERCREAM STEPS

1. Beat the unsalted butter until soft. Sift the icing sugar and add in little by little.
2. Add in the milk one teaspoon at a time. Mix in the instant sesame dessert powder. Add in the black food coloring.
3. Beat until smooth and creamy.

DECORATION STEPS

1. Place the Petal Piping Tip into a piping bag. Place the Sesame Buttercream into the piping bag and pipe onto each cupcake. Pipe out with tip position at 45 degree angle to cupcake, apply with steady pressure, pipe out icing on cupcake in an up & down motion going around to form a ruffle flower.
2. Sprinkle toasted black sesame in the middle (decorate as core of the flower).

菠蘿包蛋糕

菠蘿包是香港的特產，誕生於60年代，是港式茶餐廳的經典食物！只要加1塊奶油就變身為美味的菠蘿油，再配搭1杯又香又滑的奶茶，就是香港人最佳的「三點三」。cupcake是外國人的小點心，和菠蘿油一樣，是下午茶的主角。所以我做了這個fusion版的菠蘿包cupcake：一次吃2種下午茶點，是不是很過癮呢？

材料

蛋糕

□基本香草奶油蛋糕麵糊 1份（作法詳見19頁）

菠蘿皮

□無鹽奶油 20克 □白砂糖 27.5克 □奶粉 3.5克 □奶水 1/2湯匙
□煉乳 1/2茶匙 □蛋黃 1顆（1/2顆用於菠蘿脆皮，1/2顆用於蛋糕表面）
□泡打粉 1/2茶匙 □蘇打粉 1/4茶匙 □低筋麵粉或普通麵粉 62.5克

製作蛋糕

準備麵糊：在馬芬杯裡放入紙模，倒入預先準備好的基本香草奶油蛋糕麵糊，約2/3滿。

製作菠蘿皮

1. **打發奶油：**奶油置於室溫回軟，加入糖，打至無粉粒狀態。
2. **加入奶類和蛋黃：**依次加入奶粉、半顆蛋黃、奶水、煉乳，攪拌均勻。
3. **篩入粉類：**將低筋麵粉（或普通麵粉）、泡打粉、蘇打粉混合過篩，加入其中，攪拌成麵團。
4. **冷藏麵團：**將麵團做成長圓筒形，包上保鮮膜，放進冰箱冷藏30～40分鐘。
5. **壓成圓片狀：**從冰箱拿出麵團分6等份，再用手搓成圓球狀，然後壓平至圓片狀（直徑約2吋）。
6. **用小刀劃出花紋：**用小刀輕劃菠蘿皮表面，劃出深約0.1公分的格子紋路，注意不要切斷脆皮。
7. **烘焙：**將菠蘿皮放在蛋糕糊的上面，表面刷上蛋液，用170℃烤20～30分鐘或烤至呈金黃色。

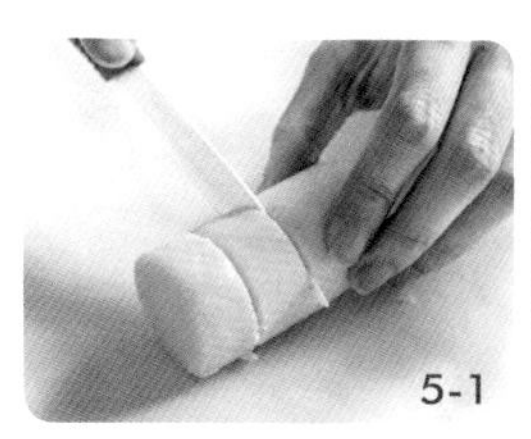

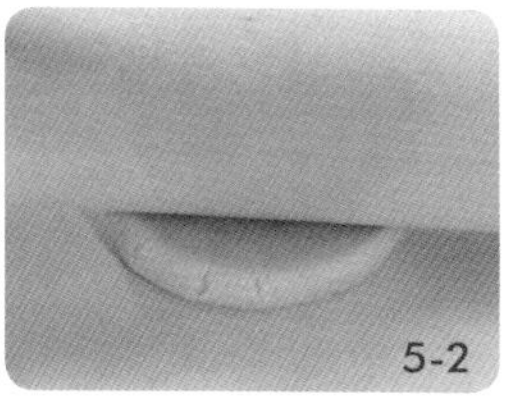

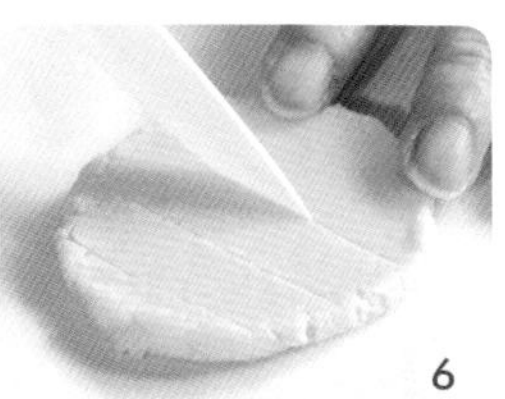

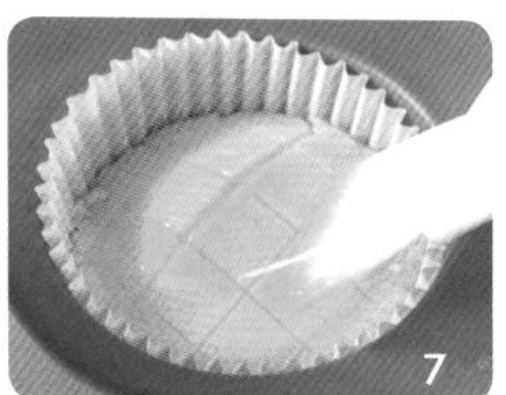

裝飾步驟（可選擇不做）

夾入厚切奶油：把蛋糕切成一半，中間夾入1塊厚切奶油片。

Pineapple Bun Cupcakes

INGREDIENTS

For Cupcake

1 portion of Basic Vanilla Cupcake Batter (please refer to page 19)

For Pineapple-Bun Topping

62.5g cake flour or plain flour
27.5g caster sugar
20g unsalted butter (at room temperature)
3.5g milk powder
1 egg yolk (1/2 for the bun topping, 1/2 for brushing)
1/2 tablespoon evaporated milk
1/2 teaspoon condensed milk
1/2 teaspoon baking powder
1/4 baking soda

PINEAPPLE CUPCAKE STEPS

Prepare 1 portion of Basic Vanilla Cupcake Batter. Pour into cupcake cups, 2/3 full.

PINEAPPLE-BUN TOPPING STEPS

1. Beat the butter until smooth. Add in the sugar until white and fluffy.
2. Add in milk powder, 1/2 egg yolk, evaporated milk, condensed milk.
3. Sift the cake flour, baking powder and baking soda and add into mixture.
4. Wrap up the dough into a long cylinder shape and refrigerate it for 30-40 mins, use cling wrap to wrap as its very sticky and wet.
5. Divide the dough into 6 even portions and roll each into a round ball. Use your palm to push each ball into a flat round disc shape.
6. Place and centre on top of each cupcake. Use a knife to draw diamond grids on top.
7. Brushed with the remaining egg yolk. Bake together with the cupcakes for 20-30 minutes or until golden brown.

FOR DECORATION (OPTIONAL)

Cut the cake into half and place in the cold butter slice.

杏仁餅蛋糕

如果要選一個代表澳門的食品，應該非杏仁餅莫屬。杏仁餅本身比較乾，有很香的杏仁味和杏仁粒。這款杏仁餅cupcake，不但杏仁味十足，而且口感綿軟，讓你可以嘗到杏仁餅的另一種風格。

材料

蛋糕

- 杏仁香精 1茶匙
- 杏仁碎粒 15克
- 基本香草奶油蛋糕麵糊 1份（作法詳見19頁）

杏仁奶油糖霜

- 杏仁香精 1/2茶匙
- 基本奶油糖霜 1份（作法詳見21頁）

裝飾

- 杏仁碎粒 適量
- 杏仁餅 適量
- 擠花袋 1個
- 開縫星齒花嘴 1個

製作蛋糕

1. **加入杏仁香精、碎粒：**將杏仁香精、杏仁碎粒放入預先準備好的基本香草奶油蛋糕麵糊中。
2. **入烤箱烘焙：**將麵糊慢慢倒入蛋糕杯，約2/3滿。用170℃烤20～30分鐘或烤至蛋糕表面呈金黃色。
3. **冷卻：**完成後，轉移到散熱架上。等蛋糕完全冷卻後才裝飾。

製作杏仁奶油糖霜

1. **加入杏仁香精：**將杏仁香精加入預先準備好的基本奶油糖霜裡，攪拌均勻。
2. **打發糖霜：**用電動攪拌器打發至無粉粒狀態。

裝飾步驟

1. **使用開縫星齒花嘴：**拿1個擠花袋，放入開縫星齒花嘴。將杏仁奶油糖霜放入擠花袋，往下推實，紮緊袋口，在蛋糕表面擠上花紋。
2. **放上杏仁餅和杏仁碎粒：**放上杏仁餅，於蛋糕表面撒上一些杏仁碎粒。

Almond Cookies Cupcakes

INGREDIENTS

For Cupcake

1 portion of Basic Vanilla Cupcake Batter (please refer to page 19)
1 teaspoon almond extract
15g almond bits

For Almond Buttercream

1 portion of Basic Buttercream (please refer to page 21)
1/2 teaspoon almond extract

For Decoration

Open Star Piping Tip
1 piping bag
almond bits (for decoration)
almond cookies

CUPCAKE STEPS

1. Add the almond essence into 1 portion of Basic Vanilla Cupcake Batter. Stir in the almond bits.
2. Pour into cupcake cups, 2/3 full. Bake for 20-30 minutes or until golden brown.
3. Transfer to cooling rack. Cool down completely before frosting.

ALMOND BUTTERCREAM STEPS

1. Add the almond extract into 1 portion of Basic Buttercream.
2. Mix until well blended.

DECORATION STEPS

1. Place the Open Star Piping Tip into a piping bag. Place the Almond Buttercream into the piping bag and pipe onto each cupcake.
2. Decorate with almond cookies. Sprinkle almond bits on top.

木糠布丁蛋糕

木糠布丁是澳門別具特色的葡式甜品。不同的餐廳會有不同的做法。木糠是指餅乾碎，而布丁是由奶油等材料組成的。我每次去澳門都會吃這道甜點，但是總是覺得好像欠缺一些東西，口感太creamy，所以我加了蛋糕去做這個木糠布丁cupcake，來填補我覺得欠缺的東西。

材料

蛋糕

□基本香草奶油蛋糕麵糊 1份（作法詳見19頁）

裝飾

□鮮奶油 250毫升 □煉乳 2湯匙

□消化餅乾 2塊 □擠花袋 1個 □開縫星齒花嘴 1個

製作步驟

1 **製作基本奶油蛋糕**：把準備好的基本香草奶油蛋糕麵糊倒入蛋糕杯，約2/3滿。用170℃烤20～30分鐘或烤至蛋糕表面呈金黃色。

2 **填入煉乳**：在每個蛋糕中間挖1個小孔，在小孔中間填滿煉乳。

3 **打發鮮奶油**：將淡奶油打發至硬性發泡，加入煉乳，拌勻。

4 **擠上花紋**：拿1個擠花袋，放入開縫星齒花嘴。將已打發的鮮奶油放入擠花袋，往下推實，紮緊袋口，在蛋糕表面擠上花紋。

5 **撒上消化餅乾碎**：最後，撒一些壓碎的消化餅乾作木糠裝飾。

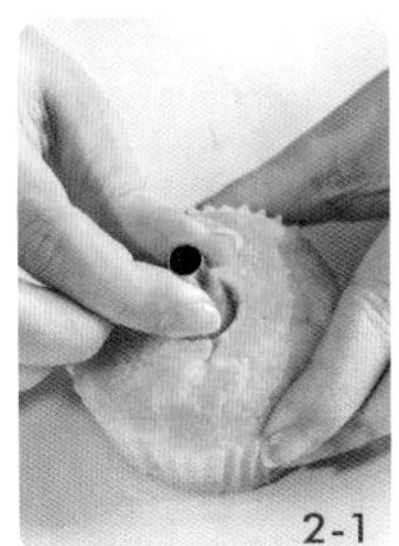
2-1

2-2

3

4

5

Serrandura Cupcakes

INGREDIENTS

For Cupcake

1 portion of Basic Vanilla Cupcake Batter (please refer to page 19)

For Decoration

Open Star Piping Tip
1 piping bag
250ml whipped cream
2 tablespoons condensed milk
2 pieces of digestive biscuits (crumbs)

STEPS

1. Pour 1 portion of Basic Vanilla Cupcake Batter into 6 cupcake cups, 2/3 full. Bake for 20-30 minutes or until golden brown.
2. Cut a hole in the middle of the cupcakes. Fill with condensed milk.
3. Whip the cream to peak. Fold in the condensed milk.
4. Place the Open Star Piping Tip into a piping bag. Place the whipped cream into the piping bag and pipe onto each cupcake.
5. Sprinkled with digestive biscuit crumbs.

臺灣
TAIWAN

由於地處熱帶及亞熱帶氣候區之交界，

臺灣的自然景觀與生態資源相當豐富。

人口以原住民族和漢族為主，

或分為原住民族、河洛人、客家人、外省人、新住民等五大族群。

因為有那麼多不同民族，使得食物也非常多元化，

各族群傳統的食物做法也大多沒有失傳。

每次到臺灣必去逛夜市，各種美食讓人吃到停不下來！

鳳梨酥、太陽餅、珍珠奶茶、炸雞、肉燥飯、三杯雞等等，

都是臺灣的招牌美食，絕對值得一試！

鳳梨蛋糕

鳳梨酥在臺灣有許多不同的品牌，味道更是五花八門，已經進化到有草莓味或芒果味，甚至有些還加入蛋黃。這款鳳梨味道的cupcake，不但有新鮮鳳梨粒和鳳梨餡，就連蛋糕的表面都做出鳳梨的樣子。不知道這個鳳梨cupcake將來會不會取代鳳梨酥，成為另一種熱門伴手禮呢？

材料

蛋糕

- □ 無鹽奶油 55克
- □ 白砂糖 50克
- □ 雞蛋 1顆
- □ 鳳梨罐頭或新鮮鳳梨切粒 30克
- □ 自發粉 100克
- □ 脱脂牛奶 40毫升
- □ 植物油 20克
- □ 鳳梨香精 1/2茶匙

鳳梨奶油糖霜

- □ 基本奶油糖霜 1份（作法詳見21頁）
- □ 黃色食用色素 1～2滴
- □ 綠色食用色素 1～2滴
- □ 鳳梨香精 1/2茶匙

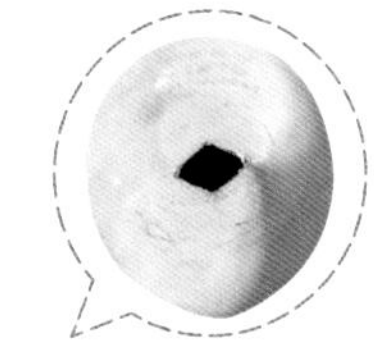

裝飾

- □ 鳳梨餡（即食餡料，烘焙店可買）
- □ 擠花袋 2個
- □ 開縫星齒花嘴
- □ 葉形花嘴

製作蛋糕

1

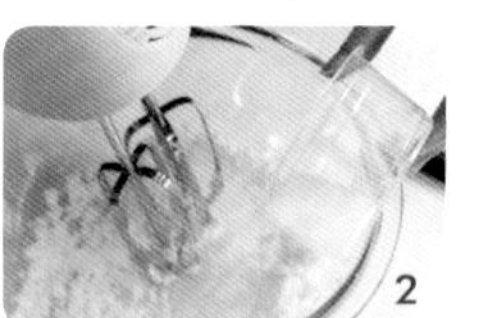
2

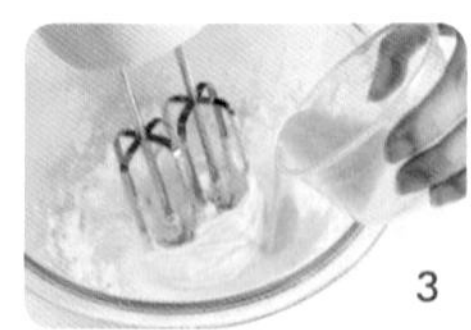
3

4

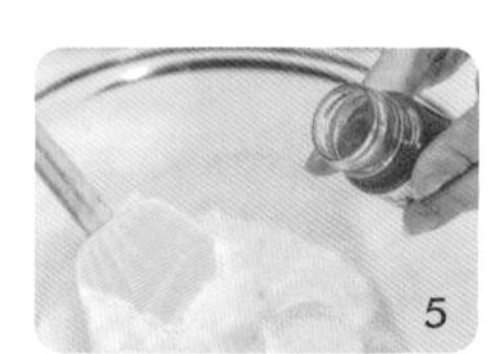
5

6

7

1 **準備**：將烤箱預熱至170℃。奶油在室溫中回軟。

2 **打發奶油**：加入白砂糖，和奶油一起打發，直到顏色變淺，呈軟滑狀。

3 **加入雞蛋**：加入打散的蛋液，均勻攪拌，直到和奶油混合。

4 **篩入粉類**：將自發粉過篩，分3次加入。用矽膠刮刀以切拌的方式快速拌勻。

5 **加入其他材料**：慢慢加入脱脂牛奶。再分別加入植物油、鳳梨香精和鳳梨罐頭或新鮮鳳梨切粒。

6 **入烤箱烘焙**：將麵糊慢慢倒入蛋糕杯，約2/3滿。用170℃烤20～30分鐘或烤至蛋糕表面呈金黃色。

7 **冷卻**：完成後，轉移到散熱架上。等蛋糕完全冷卻後才裝飾。

製作鳳梨奶油糖霜

1 **混合**：將鳳梨香精均勻拌入預先準備好的基本奶油糖霜中。

2 **加入黃色色素**：在3/4份糖霜裡加入黃色食用色素，拌勻。

3 **加入綠色色素**：在剩餘1/4份糖霜裡加入綠色食用色素，拌勻。

1

2

3

裝飾步驟

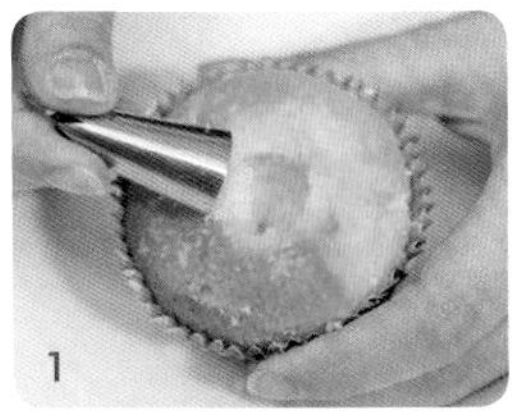

1 **填餡：**在蛋糕中間挖1個小孔，填入鳳梨餡。

2 **使用開縫星齒花嘴：**先在蛋糕表面抹上1層黃色鳳梨奶油糖霜。再拿1個擠花袋，放入開縫星齒花嘴。將黃色鳳梨奶油糖霜放入擠花袋，往下推實，紮緊袋口，在蛋糕表面擠上星形花紋。

3 **使用葉形花嘴：**再拿出1個擠花袋，放入葉形花嘴。將綠色鳳梨奶油糖霜放入擠花袋，往下推實，紮緊袋口，在蛋糕表面擠上葉形花紋。

Pineapple Cupcakes

INGREDIENTS

For Cupcake

55g unsalted butter
 (at room temperature)
50g caster sugar
1 large egg
100g self-raising flour
40ml skimmed milk
 (at room temperature)
20g vegetable oil
1/2 teaspoon pineapple essence
30g canned or fresh
 pineapple pieces

For Pineapple Buttercream

1 portion of Basic Buttercream
 (please refer to page 21)
1/2 teaspoon pineapple essence
1-2 drops of yellow food coloring
1-2 drops of green food coloring

For Decoration

Open Star Piping Tip –for the main pineapple yellow part
Leaf Piping Tip –or pineapple stem part in green color
2 piping bags
pineapple paste for filling (instant can buy from bakery shop)

CUPCAKE STEPS

1. Preheat the oven to 170 degrees Celsius.
2. Beat the butter and sugar until smooth.
3. Add in the egg and mix well.
4. Sift the self-raising flour.
5. Add in the flour little by little to the batter.
 Add in the milk little by little.
 Add in the vegetable oil and the pineapple essence.
 Stir in the crushed pineapple pieces.
6. Pour into cupcake cups, 2/3 full. Bake for 20-30 minutes or until golden brown.
7. Transfer to cooling rack.
 Cool down completely before frosting.

PINEAPPLE BUTTERCREAM ICING STEPS

1. Add the pineapple essence to 1 portion of Basic Buttercream.
2. Take 3/4 of buttercream, mix in yellow food coloring.
3. Take 1/4 remaining buttercream, mix in green food coloring.

DECORATION STEPS

1. Cut a hole in the middle of each cupcake. Fill centre with pineapple filling.
2. Place the Open Star Piping Tip into a piping bag. Place the Yellow Pineapple Buttercream into the piping bag and pipe onto each cupcake.
3. Place the Leaf Piping Tip into another piping bag.
 Place the Green Pineapple Buttercream into another piping bag and pipe the Green Pineapple Buttercream onto the top of the pineapple as the stem/leaves.

紫薯蛋糕

這個紫番薯cupcake的糖霜非常健康，因為用了紫番薯餡。這個味道的靈感來自臺灣九份，當地稱番薯為地瓜。在九份有很多地瓜製成的食品，例如地瓜酥、地瓜餅、地瓜芋圓等等，我很想把這個味道帶回香港，就做了紫薯cupcake。

材料

蛋糕

- 紫薯泥 1/2個（約100克）
- 基本香草奶油蛋糕麵糊 1份（作法詳見19頁）
- 甜紫薯乾 10克（可選擇不加）
- 紫色的食用色素 適量（可選擇不加）

紫薯奶油糖霜

- 脫脂牛奶 2～3茶匙
- 糖粉 100克
- 無鹽奶油 100克
- 紫薯 2個（約360克）

裝飾

- 紫薯乾（可選擇不加）
- 紫色糖碎
- 擠花袋1個
- 糖花
- 糖蝴蝶
- 多瓣花齒花嘴

製作紫薯泥醬

1. **紫薯煮至軟熟：**將紫薯放入小鍋中，加入熱水，以中火煮約30分鐘至完全軟綿熟透。
2. **壓成泥醬：**將水倒掉，待紫薯稍降溫後，除去外皮，用金屬湯匙背面將紫薯壓成泥醬備用。＊可用牙籤插入紫薯，測試是否熟透。

1

2

製作蛋糕

1. **混合麵糊和紫薯：**將壓成泥醬的紫薯加入準備好的基本香草奶油蛋糕麵糊中。
2. **加入其他材料：**加入紫色的食用色素和紫薯乾。
3. **入烤箱烘焙：**將麵糊慢慢倒入蛋糕杯，約2/3滿。用170℃烤20～30分鐘或烤至蛋糕表面呈金黃色。
4. **冷卻：**完成後，轉移到散熱架上。等蛋糕完全冷卻後才裝飾。

製作紫薯奶油糖霜

1. **打發奶油：**將室溫奶油打至軟綿。
2. **加入粉類：**分2～3次加入糖粉。
3. **加入其他材料：**加入紫薯泥醬，拌勻；加入脫脂牛奶。
4. **繼續打發：**打發至鬆軟順滑。

裝飾步驟

1. **使用多瓣花齒花嘴：**拿1個擠花袋，放入多瓣花齒花嘴。將紫薯奶油糖霜放入擠花袋，往下推實，紮緊袋口，在蛋糕表面擠上花紋。
2. **放上糖花裝飾：**放上糖飾花朵和蝴蝶，撒上紫色糖碎即可。

Sweet Potato Cupcakes

INGREDIENTS

For Cupcake

1 portion of Basic Vanilla Cupcake Batter (please refer to page 19)
1/2 small purple sweet potatoes (mashed, around 100g)
purple food coloring (optional)
10g sprinkle dried sweet potatoes (optional)

For Sweet Potato Buttercream

100g unsalted butter (at room temperature)
100g icing sugar
2-3 teaspoons of milk
2 small purple sweet potatoes (mashed, around 360g)

For Decoration

Closed Star Piping Tip
1 piping bag
dried sweet potato bits (optional)
suger flowers
suger butterflies
sprinkles

CUPCAKE STEPS

1. Stir in the mashed purple sweet potatoes to 1 portion of Basic Vanilla Cupcake Batter.
2. Add a drop of purple food coloring if the batter is not purple enough. Stir in the dried sweet potatoes.
3. Pour into cupcake cups, 2/3 full. Bake for 20-30 minutes or until golden brown.
4. Transfer to cooling rack. Cool down completely before frosting.

MASHED SWEET POTATO PASTE STEPS

1. Put one sweet potato into a pan and pour in hot water, enough to cover the entire sweet potato. Cook with medium heat for around 30 minutes until sweet potato becomes soften, checked with the insert of the toothpick or a small knife into the center.
2. Remove from water, let sweet potato cool down, then peel off the skin and mash it with a metal spoon.

BUTTERCREAM STEPS

1. Beat the unsalted butter until soft.
2. Add in icing sugar and beat until smooth.
3. Add in the mashed sweet potatoes. Add in the milk, one teaspoon at a time.
4. Beat until smooth and creamy.

DECORATION STEPS

1. Place the Closed Star Piping Tip into a piping bag.
2. Place 1 portion of Sweet Potato Buttercream into a piping bag and pipe onto the cupcakes.
3. Decorate with sugar flowers, butterflies and sprinkles.

珍珠奶茶蛋糕

珍珠奶茶是我最喜歡的臺灣飲料，雖然現在香港有很多地方都可以買到，不一定要飛去臺灣。不過我還是嫌不過癮，想嘗試新花樣，於是就做了這款珍珠奶茶cupcake！我用了紅茶、牛奶和黑糖香精，讓蛋糕有珍珠奶茶的香味，再加上黑珍珠，使口感更為豐富。 這個蛋糕會裝飾成珍珠奶茶的樣子，不過是1杯可以吃的珍珠奶茶哦！

材料

蛋糕

- 無鹽奶油 55克
- 黑糖 50克
- 雞蛋 1顆
- 自發粉 100克
- 脫脂牛奶 10毫升
- 植物油 25克
- 黑珍珠粒 2湯匙（約30克，烘焙店可買）
- 黑糖香精 1/2茶匙
- 紅茶茶包1包（用35毫升熱水泡15分鐘）

裝飾

- 黑珍珠粒
- 蛋白霜 1份（作法詳見22頁）
- 彩色吸管
- 擠花袋 1個
- 開縫星齒花嘴

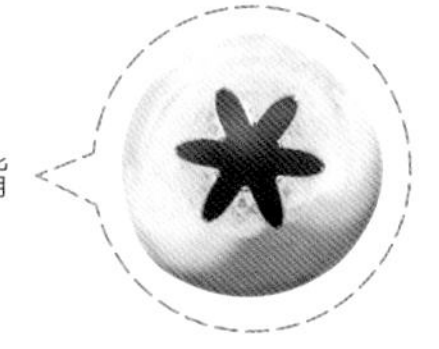

製作蛋糕

1 **準備**：將烤箱預熱至170℃。奶油在室溫中回軟。

2 **打發奶油**：加入黑糖，和奶油一起打發，直到顏色變淺，呈軟滑狀。

3 **加入雞蛋**：加入打散的蛋液，均勻攪拌，直到和奶油混合。

4 **篩入粉類**：將自發粉過篩，然後分3次加入。用矽膠刮刀以切拌的方式快速拌勻。

5 **加入其他材料**：依次加入脫脂牛奶和紅茶。加入植物油、黑糖香精和黑珍珠粒。

6 **入烤箱烘焙**：將麵糊慢慢倒入蛋糕杯，約2/3滿。用170℃烤20～30分鐘或烤至蛋糕表面呈金黃色。

7 **冷卻**：完成後，轉移到散熱架上。等蛋糕完全冷卻後才裝飾。

裝飾步驟

1 **使用開縫星齒花嘴**：拿1個擠花袋，放入開縫星齒花嘴。將蛋白霜放入擠花袋，往下推實，紮緊袋口，在蛋糕表面擠上花紋。

2 **放上珍珠粒和吸管**：放一些黑珍珠粒在蛋白霜上面，再插1支吸管作裝飾。

小技巧 黑珍珠怎麼煮才更Q彈爽滑？

1.控制水量→將黑珍珠放入小鍋中，加入約黑珍珠4倍的水量，以中火煮到黑珍珠浮上面，改小火煮10分鐘後熄火。

2.熄火蓋上蓋子→蓋上蓋子煮10分鐘。

3.冷卻→黑珍珠以飲用冷水沖過後，冷卻備用。

★在烘焙店可買到真空脫水包裝的黑色珍珠粉圓。

1-1

1-2

3

Bubble Tea Cupcakes

INGREDIENTS

For Cupcake

55g unsalted butter (at room temperature)
50g light brown sugar
1 egg
100g self-raising flour
1 teabag black tea (soaked in 35ml hot water for 15 mins)
10ml skimmed milk
25g vegetable oil
1/2 teaspoon black sugar essence
2 tablespoons black bubbles (30g) (buy in bakery shop)

For Decoration

Open Star Piping Tip
1 piping bag
1 portion of Meringue Frosting (please refer to page 22)
extra black bubbles
drinking straws

CUPCAKE STEPS

1.Preheat oven 170 celcius.
2.Beat butter and sugar until smooth.
3.Add in beaten egg and mix well.
4.Sift the self-raising flour. Add in the flour little by little to the batter.
5.Add in the milk and black tea little by little. Add in the vegetable oil and black sugar essence. Stir in black bubbles.
6.Pour into cupcake cups, 2/3 full. Bake for 20-30 minutes or until golden brown.
7.Transfer to cooling rack. Cool down completely before frosting.

DECORATION STEPS

1.Place the Open Star Piping Tip into a piping bag. Place the Meringue Frosting into the piping bag and pipe onto the cupcake.
2.Topped with black bubbles. Decorate with a drinking straw.

TIPS : Preparation Of The Black Bubbles

1.Put dried tapioca balls into a pan, put in water (tapioca weight x 4 times) . Bring to boil until all tapioca balls starts floating on the surface of the water. Turn down to medium heat and continue to cook for 10 minutes.
2.Turn the heat off and cover the pan with pan lid for another 10 minutes.
3.Wash the hot tapioca balls with drinking cold water (cold from fridge) . Set aside for later use.

* You can buy Tapioca Balls in baking supply stores called Instant Dried Black Tapioca Ball.

黑糖薑母茶蛋糕

薑母茶一般是由老薑及黑糖或紅糖所製成。以中醫的角度，薑母茶具有去風寒、開脾胃的功效，所以很多人現在去臺灣都會買這個當伴手禮。我當然不會錯過這麼受歡迎的口味！這款黑糖薑母茶cupcake，表面選用另外一個我很喜歡吃的薑餅作裝飾，相信是最有特色的自製臺灣伴手禮！

材料

蛋糕

- 鹽 適量
- 薑粉 1湯匙（先溶於15毫升熱水）
- 黑糖薑母茶粉 1/2茶匙
- 基本香草奶油蛋糕麵糊 1份（作法詳見19頁）

薑味奶油糖霜

- 薑粉 20克
- 糖粉 180克
- 無鹽奶油 100克
- 脫脂牛奶 2～3茶匙

裝飾

- 薑餅 1塊
- 擠花袋 1個
- 開縫星齒花嘴

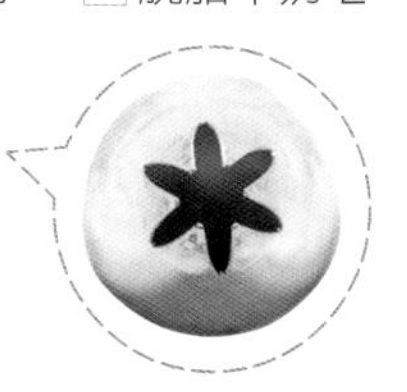

製作蛋糕

1. **混合麵糊和其他材料**：將薑粉和黑糖薑母茶粉溶液加入預先準備好的基本香草奶油蛋糕麵糊，拌勻，再加少許鹽。
2. **入烤箱烘焙**：將麵糊倒入蛋糕杯，約2/3滿。用170℃烤20～30分鐘或烤至蛋糕面呈金黃色。
3. **冷卻**：完成後，轉移到散熱架上。等蛋糕完全冷卻後才裝飾。

製作薑味奶油糖霜

1. **打發奶油**：將室溫奶油打發至軟綿。
2. **加入糖粉、薑粉**：分5～6次加入糖粉、薑粉。
3. **加入牛奶**：加入脫脂牛奶。
4. **繼續打發**：繼續打發至鬆軟順滑。

裝飾步驟

1. **使用開縫星齒花嘴**：拿1個擠花袋，放入開縫星齒花嘴。將薑味奶油糖霜放入擠花袋，往下推實，紮緊袋口，在蛋糕表面擠上花紋。
2. **放上薑餅**：放上1塊薑餅作裝飾。

Ginger Cupcakes

INGREDIENTS

For Cupcake

1 portion of Basic Vanilla Cupcake Batter (please refer to page 19)
1 tablespoon ginger powder (dissolved in 15ml hot water)
1/2 teaspoon brown sugar ginger tea powder
pinch of salt

For Ginger Buttercream

100g unsalted butter (at room temperature)
180g icing sugar
20g ginger powder
2-3 teaspoon of skimmed milk

For Decoration

Open Star Piping Tip
1 piping bag
ginger cookie

CUPCAKE STEPS

1.Add in the ginger powder liquid and brown sugar ginger tea powder into Basic Vanilla Cupcake Batter. Add in a pinch of salt.
2.Pour into cupcake cups, 2/3 full. Bake for 20-30 minutes or until golden brown.
3.Transfer to cooling rack. Cool down completely before frosting.

BUTTERCREAM STEPS

1.Beat the unsalted butter until soft.
2.Add in sugar and ginger powder and mix well.
3.Add in the milk, 1 teaspoon at a time.
4.Beat until smooth and creamy.

DECORATION STEPS

1.Place the Open Star Piping Tip into a piping bag. Place the Ginger Buttercream into the piping bag and pipe on to the cupcake.
2.Topped with ginger cookie.

日本

日本人對食物的外觀和烹調方法非常要求。
大家有沒有留意到日本人吃東西前，會將雙手合十，
然後說「Itadakimasu」，這句話可以譯作「我不客氣了」。
早在1896年，日本著名養生學家石冢左玄就已經在其著作
《食物養生法》提出「人類作為食物鏈的其中一環，
對進食應抱持感恩態度」之想法。
正因如此，他們在處理和烹調食物的時候都會很用心。
日本的食物包裝也非常精緻，
我每次買回東西都不捨得拆開來吃。
這一章所製作的cupcake，除了吸收日本美食的經典口味外，
也會在裝飾上體現細膩的和風！

綠茶蛋糕

在日本用餐，大概每間餐廳都會給你1杯綠茶，特別是吃完甜點或是一些炸物之後，喝1杯綠茶可以解膩。除此之外，綠茶還有降膽固醇、消脂、防止蛀牙和抗氧化等功效，確實有很多好處。所以日本人每日都會喝綠茶，而且很多甜品都會使用綠茶製作。

材料

蛋糕

- 無鹽奶油 45克
- 白砂糖 70克
- 雞蛋 1顆
- 自發粉 90克
- 植物油 20克
- 脫脂牛奶 30毫升
- 綠茶粉 2茶匙
- 綠茶 15毫升（先溶於15毫升熱水內）

綠茶法式蛋白奶油霜

- 綠茶粉 2湯匙
- 法式蛋白奶油霜 1份（作法詳見23頁）

裝飾

- 綠茶巧克力手指餅
- 綠茶粉 適量
- 擠花袋 1個
- 開縫星齒花嘴 1個

製作蛋糕

2

5-2

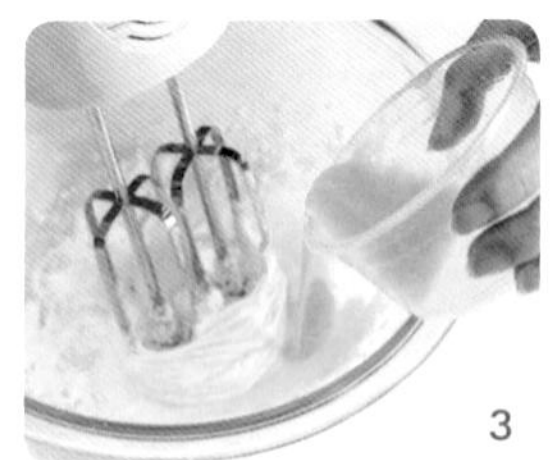

3

5-3

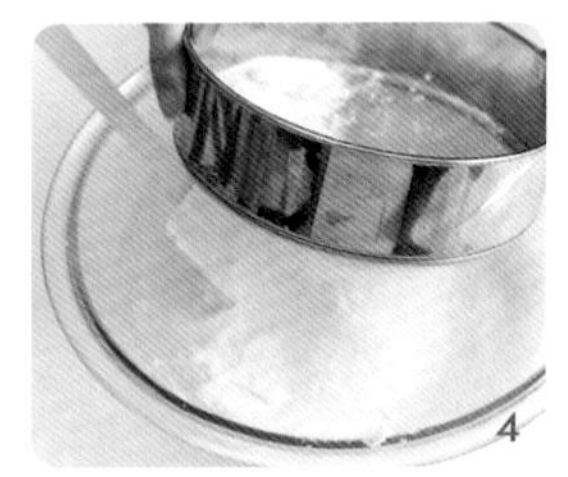

4

6

5-1

7

1 **準備：**將烤箱預熱至170℃。奶油在室溫中回軟。

2 **打發奶油：**加入白砂糖，和奶油一起打發，直到顏色變淺，呈軟滑狀。

3 **加入雞蛋：**加入打散的蛋液，混合攪拌，直到和奶油融合。

4 **篩入粉類：**將自發粉過篩，然後分3次加入。用矽膠刮刀以切拌的方式快速拌勻。

5 **加入其他材料：**依次加入脫脂牛奶、綠茶、植物油和綠茶粉。

6 **入烤箱烘焙：**將麵糊慢慢倒入蛋糕杯，約2/3滿。用170℃烤20～30分鐘或烤至蛋糕表面呈金黃色。

7 **冷卻：**完成後移到散熱架上。等蛋糕完全冷卻後才裝飾。

製作綠茶法式蛋白奶油霜

1 **拌入綠茶粉**：將綠茶粉均勻拌入預先準備好的法式蛋白奶油霜內。

2 **攪拌均勻**：用電動攪拌器打發至無粉粒狀態。

裝飾步驟

1 **使用開縫星齒花嘴**：拿1個擠花袋，放入開縫星齒花嘴。將綠茶蛋白奶油霜放入擠花袋，往下推實，紮緊袋口，在蛋糕表面擠上花紋。

2 **用綠茶粉和綠茶巧克力手指餅裝飾**：在奶油霜上撒上綠茶粉，再放上2根綠茶巧克力手指餅作裝飾。

Green Tea Cupcakes

INGREDIENTS

For Cupcake

45g unsalted butter
（at room temperature）
70g caster sugar
1 large egg
90g self-raising flour
30ml skimmed milk
15 ml green tea
（green tea powder dissolved in 15ml hot water）
20g vegetable oil
2 teaspoon green tea powder

For Green Tea French Meringue Buttercream

1 portion of French Meringue Buttercream
（please refer to page 23）
2 tablespoons green tea powder

For Decoration

Open Star Piping Tip
1 piping bag
extra green tea powder
（to sprinkle on top）
green tea chocolate coated biscuit sticks

CUPCAKE STEPS

1.Preheat the oven to 170 degrees Celsius.
2.Beat the butter and sugar until smooth.
3.Add in the egg and mix well.
4.Sift the self-raising flour. Add in the flour little by little to the batter.
5.Add in the milk, green tea（water form）, vegetable oil little by little. Stir in green tea powder.
6.Pour into cupcake cups, 2/3 full. Bake for 20-30 minutes or until golden brown.
7.Transfer to cooling rack. Cool down completely before frosting.

GREEN TEA FRENCH MERINGUE BUTTERCREAM STEPS

1.Add the green tea powder into 1 portion of French meringue buttercream.
2.Beat until well blended and smooth.

DECORATION STEPS

1.Place the Open Star Piping Tip into a piping bag. Place the green tea powder into the Meringue Buttercream. Sprinkle with extra green tea powder on top of cupcake.
2.Decorate with 2 green tea chocolate coated biscuit sticks.

巨峰葡萄蛋糕

葡萄當中，我最喜歡巨峰葡萄，因為它有獨特的味道。秋天直至11月是巨峰葡萄的收成期，以有水果王國美譽的日本山梨縣為最佳。若你喜歡巨峰，一久園可以容許你在園內任意吃上40分鐘，費用只需大約130港元（約550元臺幣）。但如果沒機會去日本，香港也有很多地方可以買到。這個cupcake加入了我最喜歡的巨峰葡萄粒，裝飾部分也別具新意，大家不妨動手做做看！

材料

蛋糕

- 無鹽奶油 45克
- 白砂糖 70克
- 自發粉 90克
- 脫脂牛奶 30毫升
- 雞蛋 1顆
- 新鮮巨峰葡萄汁 10毫升
- 植物油 20克
- 新鮮巨峰葡萄粒 20克

巨峰葡萄法式蛋白奶油霜

- 紫色食用色素 1滴
- 葡萄香精 1/2茶匙
- 法式蛋白奶油霜 1份（作法詳見23頁）

裝飾

- 葡萄軟糖
- 綠色葉形軟糖或葡萄綠色莖部分

製作蛋糕

5

1. **準備：**將烤箱預熱至170℃。奶油在室溫中回軟。
2. **打發奶油：**加入白砂糖，和奶油一起打發，直到顏色變淺，呈軟滑狀。
3. **加入雞蛋：**加入打散的蛋液，混合攪拌，直到和奶油融合。
4. **篩入粉類：**將自發粉過篩，然後分3次加入。用矽膠刮刀以切拌的方式快速拌勻。
5. **加入其他材料：**依次加入脫脂牛奶、植物油、新鮮巨峰葡萄汁和新鮮巨峰葡萄粒。
6. **入烤箱烘焙：**將麵糊慢慢倒入蛋糕杯，約2/3滿。用170℃烤20～30分鐘或烤至蛋糕表面呈金黃色。
7. **冷卻：**完成後移到散熱架上。等蛋糕完全冷卻後才裝飾。

製作巨峰葡萄法式蛋白奶油霜

1. **加入色素和葡萄香精：**將紫色食用色素和葡萄香精拌勻，加入預先準備好的法式蛋白奶油霜。
2. **打發奶油霜：**用電動攪拌器打發至順滑。

裝飾步驟

2

3

1 **抹平奶油霜：**用抹刀或小湯匙將奶油霜抹平在蛋糕表面。

2 **放上綠色莖：**拿1塊綠色軟糖，或剪下葡萄綠色莖部分，放到蛋糕表面，裝飾成葡萄莖部。

3 **放上葡萄糖果：**放葡萄糖果在奶油霜上面，排列成1串葡萄的樣子。

Grape Cupcakes

INGREDIENTS

For Cupcake
45g unsalted butter (at room temperature)
70g caster sugar
1 large egg
90g self-raising flour
30ml skimmed milk
20g vegetable oil
10ml fresh grape juice
20g chopped grape pieces

For Grape French Meringue Buttercream
1 portion of French Meringue Buttercream (please refer to page 23)
1 drop of purple food coloring
1/2 teaspoon grape essence

For Decoration
round or oval shape grape sugar coated soft candies
oval shape green sugar coated soft candies (or stem of the grapes)

CUPCAKE STEPS

1.Preheat the oven to 170 degrees Celcius.
2.Beat the butter and sugar until smooth.
3.Add in the egg and mix well.
4.Sift the self-raising flour. Add in the flour little by little into the batter.
5.Add in the milk little by little. Add the vegetable oil. Add in fresh grape juice. Add in grapes.
6.Pour into cupcake cups, 2/3 full. Bake for 20-30 minutes or until golden brown.
7.Transfer to cooling rack. Cool down completely before frosting.

GRAPE FRENCH MERINGUE BUTTERCREAM STEPS

1.Add in the grape essence and purple food coloring into 1 portion of French Meringue Buttercream.
2.Beat until well blended.

DECORATION STEPS

1.Spread the Grape French Meringue Buttercream onto each cupcake using a spatula or a spoon.
2.Place oval shape green sugar coated soft candies on each cupcake or cut and trim the stems of the grapes into short mini stems and attached to each cupcake.
3.Place the grape candies on top of each cupcake and arrange like a bunch of grapes.

蜂蜜蛋糕

蜂蜜蛋糕通常都是長條形，切開幾片吃，有濃厚的蜂蜜味道，質感也是軟綿綿的，不過外形通常比較簡單。所以我決定要做個蜂蜜cupcake，彌補蜂蜜蛋糕外形上的不足。蜂蜜當然是來自於蜜蜂，所以蛋糕表面用蜂蜜奶油霜擠成蜂巢的樣子，再加上小蜜蜂作為裝飾，希望大家可以試試這個與眾不同的蜂蜜蛋糕！

材料

蛋糕

- 無鹽奶油 55克
- 白砂糖 50克
- 雞蛋 1顆
- 自發粉 100克
- 脫脂牛奶 40毫升
- 植物油 25克
- 蜂蜜 20克

蜂蜜法式蛋白奶油霜

- 法式蛋白奶油霜 1份（作法詳見23頁）
- 蜂蜜 2湯匙
- 黃色食用色素 1～2滴

裝飾

- 擠花袋 1個
- 蜜蜂裝飾糖或橢圓形黃色糖果
- 圓口花嘴
- 無需調溫白巧克力
- 無需調溫黑巧克力

製作蛋糕

1 **準備**：將烤箱預熱至170℃。奶油在室溫中回軟。

2 **打發奶油**：加白砂糖，和奶油一起打發，直到顏色變淺，呈軟滑狀。

3 **加入雞蛋**：加入打散的蛋液，均勻攪拌，直到和奶油混合。

4 **篩入粉類**：自發粉過篩，然後分3次加入。用矽膠刮刀以切拌的方式快速拌勻。

5 **加入其他材料**：依次加入脫脂牛奶、植物油和蜂蜜。

6 **入烤箱烘焙**：將麵糊慢慢倒入蛋糕杯，約2/3滿。用170℃烤20～30分鐘或烤至蛋糕表面呈金黃色。

7 **冷卻**：完成後，轉移到散熱架上。等蛋糕完全冷卻後才裝飾。

製作蜂蜜法式蛋白奶油霜

1 **拌入蜂蜜**：將蜂蜜均勻拌入預先準備好的法式蛋白奶油霜內。

2 **加入黃色色素**：加入黃色食用色素。用抹刀將顏色均勻地與奶油霜混合。

裝飾步驟

1 **使用圓口花嘴**：拿1個擠花袋，放入圓口花嘴。將蜂蜜法式蛋白奶油霜放入擠花袋，往下推實，紮緊袋口，在蛋糕表面擠上蜂巢花紋。

2 **放上蜜蜂裝飾**：放上蜜蜂裝飾糖或自製蜜蜂裝飾。

1

小技巧 自製蜜蜂裝飾

1.巧克力隔水加熱→分2個小碗，將無需調溫白巧克力、無需調溫黑巧克力隔水加熱。

2.畫出蜜蜂→將黑、白巧克力溶液分別放進兩個擠花袋，在橢圓形黃色糖果上面畫出蜜蜂形態。待巧克力變硬後才放上蛋糕表面作裝飾。

1

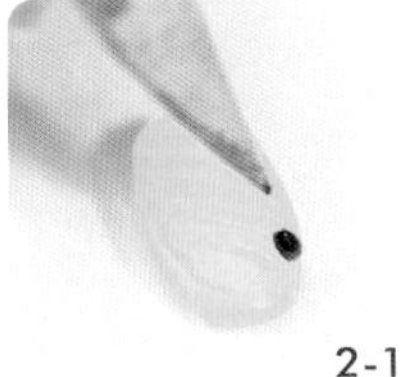
2-1

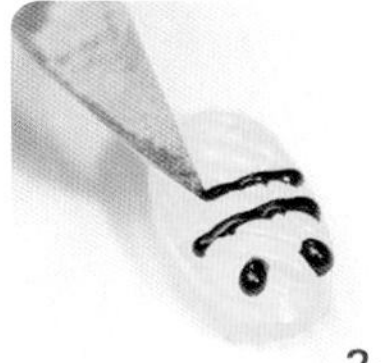
2-2

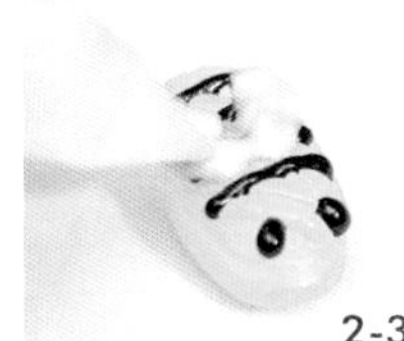
2-3

Honey Cupcakes

INGREDIENTS

For Cupcake

55g unsalted butter (at room temperature)
50g caster sugar
1 large egg
100g self-raising flour
40 ml skimmed milk
25g vegetable oil
20g honey

For Honey French Meringue Buttercream

1 portion of French Meringue Buttercream (please refer to page 23)
2 teaspoons honey
1-2 drops of yellow food coloring

For Decoration

Round Piping Tip
1 piping bag
decorative bees or yellow oval shape candies
white chocolate (for hand drawn DIY decoration)
dark chocolate (for hand drawn DIY decoration)

CUPCAKE STEPS

1.Preheat the oven to 170 degrees Celsius.
2.Beat the butter and sugar until smooth.
3.Add in the egg and mix well.
4.Sift the self-raising flour.
5.Add in the flour little by little to the batter. Add in the milk little by little. Add in the vegetable oil and the honey.
6.Pour into cupcake cups, 2/3 full. Bake for 20-30 minutes or until golden brown.
7.Transfer to cooling rack. Cool down completely before frosting.

HONEY FRENCH MERINGUE BUTTERCREAM STEPS

1.Stir the Honey into the French Meringue Buttercream.
2.Add 1-2 drops of yellow food coloring. Mix until smooth and creamy.

DECORATION STEPS

1.Place the Round Piping Tip into a piping bag. Place the Honey French Meringue Buttercream into the piping bag and pipe onto each cupcake in a swirl motion.
2.Decorate with sugar bees on top of the buttercream.

TIPS : Make Your Own DIY. Yellow Candy Bees With The Use Of Dark & White Chocolate To Hand Drawn Details.

1.Put melted chocolate in a piping bag.
2.Draw the bee outlines and details on top of oval shape candy.

紅豆蛋糕

小時候看卡通片裏面的卡通人物吃東西，是不是想一起吃呢？我印象比較深刻的是哆啦A夢的銅鑼燒。其實日本人很喜歡用紅豆來做不同的和菓子和甜品。這個紅豆cupcake裡面，除了紅豆之外，還加入了紅豆味道的豆奶，裝飾方面用了迷你版銅鑼燒，讓大家重拾有哆啦A夢相伴的童年回憶！

材料

蛋糕

- 無鹽奶油 55克
- 白砂糖 50克
- 雞蛋 1顆
- 自發粉 100克
- 香草精 1/2茶匙
- 紅豆餡 1罐
- 植物油 25克
- 紅豆味豆奶或脱脂牛奶 40毫升

裝飾

- 日本和服娃娃紙牌裝飾
- 擠花袋 1個
- 多瓣花齒花嘴
- 鮮奶油 200毫升
- 紅豆餡
- 迷你銅鑼燒 3個（日本零食店可買到）

製作蛋糕

1 **準備**：將烤箱預熱至170℃。奶油在室溫中回軟。

2 **打發奶油**：加入白砂糖，和奶油一起打發，直到顏色變淺，呈軟滑狀。

3 **加入雞蛋**：加入打散的蛋液，均勻攪拌，直到和奶油混合。

4 **篩入粉類**：自發粉過篩，然後分3次加入。用矽膠刮刀以切拌的方式快速拌勻。

5 **加入其他材料**：分次加入紅豆味豆奶或牛奶、植物油和香草精。

6 **入烤箱烘焙**：將麵糊慢慢倒入蛋糕杯，約2/3滿。用170℃烤20～30分鐘或烤至蛋糕表面呈金黃色。

7 **冷卻**：完成後，轉移到散熱架上。等蛋糕完全冷卻後才裝飾。

8 **填餡**：在每個蛋糕中間開1個小孔。將紅豆餡填入小孔。

裝飾步驟

1 **打發鮮奶油**：將鮮奶油打發至硬性發泡。

2 **使用多瓣花齒花嘴**：拿1個擠花袋，放入多瓣花齒花嘴。將鮮奶油放入擠花袋，往下推實，紮緊袋口，在蛋糕表面擠上花紋。

3 **加上紅豆餡和銅鑼燒**： 放一點點紅豆餡在鮮奶油的中間。放半個銅鑼燒在鮮奶油上面。

4 **插上日本和服娃娃紙牌**：最後，插1個日本和服娃娃紙牌裝飾。

Red Bean Cupcakes

INGREDIENTS

For Cupcake

55g unsalted butter (at room temperature)
50g caster sugar
1 large egg
100g self-raising flour
40ml red bean soy milk or skimmed milk
25g vegetable oil
1/4 teaspoon vanilla essence
1 can red bean paste

For Decoration

Closed Star Piping Tip
1 piping bag
200ml whipped cream
red bean paste
3 mini red bean buns (can buy in japan snacks shop)
Japanese doll paper toppers

CUPCAKE STEPS

1. Preheat the oven to 170 degrees Celsius.
2. Beat the butter and sugar until smooth.
3. Add in the egg and mix well.
4. Sift the self-raising flour. Add in the flour little by little to the batter.
5. Add in the red bean soya milk little by little. Add in the vegetable oil and vanilla essence.
6. Pour into cupcake cups, 2/3 full. Bake for 20-30 minutes or until golden brown.
7. Transfer to cooling rack. Cool down completely before frosting.
8. Cut a hole in the middle of the cupcakes. Fill with red bean paste.

DECORATION STEPS

1. Whip the cream to peak.
2. Place the Closed Star Piping Tip into a piping bag. Place the Whipped Cream into the piping bag and pipe onto each cupcake.
3. Place small amounts of red bean paste in centre of each cupcake.
4. Decorate with half piece of red bean bun. Decorate with Japanese doll paper topper.

新加坡・泰國

SINGAPORE
THAILAND

新加坡和泰國給人一種度假的感覺：

陽光與海灘，無拘無束，很悠閒。

上街的時候不用刻意打扮，T恤、短褲加一雙拖鞋就可以。

熱帶的天氣和沿海的地理位置，造就了當地新鮮的食材，

許多菜式都會加入在地常見的植物：

比如泰國的雞飯和蕉葉，新加坡的海南雞飯和班蘭葉。

兩地新鮮的水果、香草、海鮮也是遊客必嘗的美食！

芒果蛋糕

芒果味蛋糕很常見，而且一定是用新鮮芒果來填滿蛋糕的頂部。除了蛋糕外，還有芒果塔、芒果捲、芒果派，但是好像很少看到芒果味的杯子蛋糕，所以我做了這款杯子蛋糕。選用新鮮芒果肉做成1朵芒果花，用來送給女孩子最合適不過！但是記得要選擇比較甜的芒果哦！

材料

蛋糕

- 無鹽奶油 55克
- 白砂糖 50克
- 雞蛋 1顆
- 自發粉 100克
- 新鮮芒果汁 30毫升
- 脫脂牛奶 10毫升
- 植物油 25克
- 芒果香精 1/2茶匙
- 芒果乾 15克（可選擇不加）

裝飾

- 鮮奶油 100毫升
- 新鮮芒果切片 2～3片

製作蛋糕

1 **準備：**將烤箱預熱至170℃。奶油在室溫中回軟。

2 **打發奶油：**加入白砂糖，和奶油一起打發，直到顏色變淺，呈軟滑狀。

3 **加入雞蛋：**加入打散的蛋液，均勻攪拌，直到和奶油混合。

4 **篩入粉類：**將自發粉過篩，然後分3次加入。用矽膠刮刀以切拌的方式快速拌勻。

5 **加入其他材料：**依序加入芒果汁和牛奶。再加入植物油，加入芒果香精和芒果乾碎。

6 **入烤箱烘焙：**將麵糊慢慢倒入蛋糕杯，約2/3滿。用170℃烤20～30分鐘或烤至蛋糕表面呈金黃色。

7 **冷卻：**完成後，轉移到散熱架上。等蛋糕完全冷卻後才裝飾。

小技巧 使用甜芒果來製作，味道會更好！

裝飾步驟

1 **打發鮮奶油**：將鮮奶油打發至硬性發泡。

2 **塗抹蛋糕**：將鮮奶油塗抹在蛋糕上面。

3 **芒果切片**：芒果切成花瓣形小片，用紙巾輕壓，吸乾水分。

4 **擺面**：將芒果切片沿蛋糕中心開始向外圍繞，做成花形。

1

2

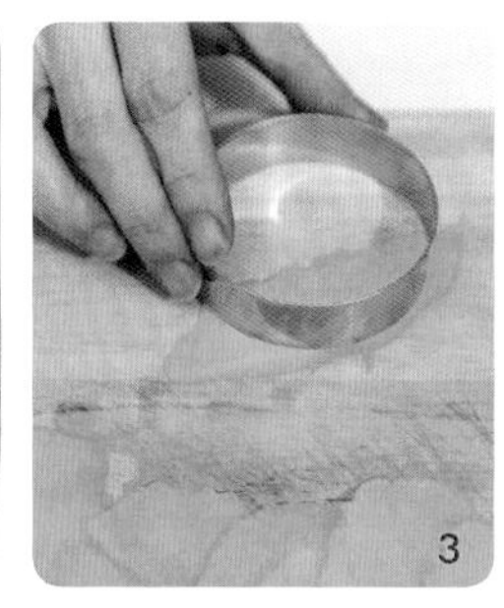
3

4

Mango Cupcakes

INGREDIENTS

For Cupcake
55g unsalted butter
(at room temperature)
50g caster sugar
1 large egg
100g self-raising flour
30ml fresh mango juice
10ml skimmed milk
25g vegetable oil
1/2 teaspoon mango essence
15g dried mango bits (optional)

For Decoration
100ml whipped cream
2-3 fresh mangos
(cut into slices)

CUPCAKE STEPS

1.Preheat the oven to 170 degrees Celsius.
2.Beat the butter and sugar until smooth.
3.Add in the egg and mix well.
4.Sift the self-raising flour. Add in the flour little by little to the batter.
5.Add in the fresh mango juice and milk little by little. Add in the vegetable oil. Add in the mango essence and dried mangoes
6.Pour into cupcake cups, 2/3 full.
Bake for 20-30 minutes or until golden brown.
7.Transfer to cooling rack. Cool down completely before frosting.

DECORATION STEPS

1.Whip the cream to peak.
2.Frost the top of the cupcake with thin layer of whipped cream.
3.Slice the mangos into small pieces.
4.Arrange mango slices around cupcake in shape of flower petals.

TIPS : **Use Sweeter Mangos.**

椰子蛋糕

一提到椰子，就會想到海島上的陽光與海灘。作為熱帶島嶼的特產，椰子不僅美味，作法也很多樣，可以製成椰子油、椰汁、椰子糖……做這個味道的cupcake，我用了椰子糖、椰子肉、椰奶和椰絲，以確保cupcake充滿濃烈的椰子味。

材料

蛋 糕

- □ 無鹽奶油 55克
- □ 白砂糖 50克
- □ 雞蛋 1顆
- □ 自發粉 100克
- □ 椰奶 40毫升
- □ 植物油 25克
- □ 椰絲碎 20克

椰子奶油糖霜

- □ 無鹽奶油 100克
- □ 糖粉 200克
- □ 椰奶 2～3茶匙
- □ 椰絲碎 10克

裝 飾

- □ 擠花袋 1個
- □ 烤椰絲
- □ 麥提莎巧克力 3粒
- □ 椰子糖漿
- □ 裝飾小紙傘 6個
- □ 多瓣花齒花嘴

製作椰子奶油糖霜

1 | **打發奶油：**將室溫奶油打至軟綿，分6～7次加入糖粉。

2 | **加入椰奶和椰絲：**加入椰奶。加入椰絲碎拌勻。

3 | **繼續打發：**用電動攪拌器打發至鬆軟順滑。

製作蛋糕

1 **準備**：將烤箱預熱至170℃。奶油在室溫中回軟。

2 **打發奶油**：加入白砂糖，和奶油一起打發，直到顏色變淺，呈軟滑狀。

3 **加入雞蛋**：加入打散的蛋液，均勻攪拌，直到和奶油混合。

4 **篩入粉類**：自發粉過篩，然後分3次加入。用矽膠刮刀以切拌的方式快速拌勻。

5 **加入其他材料**：慢慢加入椰奶。加入植物油和椰絲碎。

6 **入烤箱烘焙**：將麵糊倒入蛋糕杯，約2/3滿。用170℃烤20～30分鐘或烤至蛋糕面呈金黃色。

7 **冷卻**：完成後，轉移到散熱架上。等蛋糕完全冷卻後才裝飾。

裝飾步驟

1 **使用多瓣花齒花嘴**：拿1個擠花袋，放入多瓣花齒花嘴。將椰子奶油糖霜放入擠花袋，往下推實，紮緊袋口，在蛋糕表面擠上花紋。

2 **加上椰絲和椰子糖漿**：撒上烤椰絲，淋上一些椰子糖漿。

3 **放上巧克力裝飾**：將半粒麥提莎巧克力裝飾成椰子模樣，放在蛋糕表面上。

4 **插上小紙傘**：插1把小紙傘作裝飾。

Coconut Cupcakes

INGREDIENTS

For Cupcake
55g unsalted butter
(at room temperature)
50g caster sugar
1 large egg
100g self-raising flour
40ml coconut milk
25g vegetable oil
20g desiccated coconut
or coconut shreds

For Coconut Buttercream
100g unsalted butter
(at room temperature)
200g icing sugar
2-3 teaspoons coconut milk
10g desiccated coconut

For Decoration
Closed Star Piping Tip
1 piping bag
toasted coconut
coconut syrup
6 small party paper umbrella
3 pieces of Maltesers
chocolate (cut into halves)

CUPCAKE STEPS

1.Preheat the oven to 170 degrees Celsius.
2.Beat the butter and sugar until smooth.
3.Add in the egg and mix well.
4.Sift the self-raising flour. Add in the flour little by little to the batter.
5.Add in the coconut milk little by little. Add in the vegetable oil. Stir in the desiccated coconut or coconut shreds.
6.Pour into cupcake cups, 2/3 full. Bake for 20-30 minutes or until golden brown.
7.Transfer to cooling rack. Cool down completely before frosting.

COCONUT BUTTERCREAM STEPS

1.Beat the unsalted butter until soft. Sift the icing sugar and add in 6-7 times.
2.Add in the coconut milk, one teaspoon at a time. Stir in the desiccated coconut.
3.Beat until smooth and creamy.

DECORATION STEPS

1.Place the Closed Star Piping Tip into a piping bag. Place the Coconut Buttercream into the piping bag and pipe onto each cupcake.
2.Sprinkled with Toasted Coconut. Drizzle some Coconut Syrup on each cupcake.
3.Decorate with 1/2 a Maltesers chocolate.
4.Decorate with a party paper umbrella.

咖椰蛋糕

咖椰吐司是我去新加坡必吃的美味！我覺得咖椰介於花生醬和蜜糖之間，後兩者常常被用於製作甜點，所以我想咖椰或許是做cupcake的好材料！大家可以試試這份食譜，這樣以後除了用咖椰做吐司之外，還有另一項選擇！

材料

蛋糕

- 無鹽奶油 55克
- 白砂糖 50克
- 雞蛋 1顆
- 自發粉 100克
- 椰奶 40毫升
- 植物油 25克
- 咖椰醬 2湯匙

咖椰奶油糖霜

- 無鹽奶油 100克
- 咖椰醬 2茶匙
- 糖粉 180克
- 椰奶 2～3茶匙

裝飾

- 咖椰醬
- 迷你吐司 6塊
- 擠花袋 1個
- 開縫法式花嘴

製作蛋糕

1. **準備：**將烤箱預熱至170℃。奶油在室溫中回軟。
2. **打發奶油：**加入白砂糖，和奶油一起打發，直到顏色變淺，呈軟滑狀。
3. **加入雞蛋：**加入打散的蛋液，均勻攪拌，直到和奶油混合。
4. **篩入粉類：**自發粉過篩，然後分3次加入。用矽膠刮刀以切拌的方式快速拌勻。
5. **加入其他材料：**依序加入椰奶、咖椰醬和植物油。
6. **入烤箱烘焙：**將麵糊倒入蛋糕杯，約2/3滿。用170℃烤20～30分鐘或烤至蛋糕表面呈金黃色。
7. **冷卻：**完成後，轉移到散熱架上。等蛋糕完全冷卻後才裝飾。

製作咖椰奶油糖霜

1. **打發奶油：**將室溫奶油打至軟綿，分6～7次加入糖粉。
2. **加入椰奶：**加入椰奶。
3. **加入咖椰醬：**加入咖椰醬攪拌均勻。
4. **繼續打發：**打發至鬆軟、無粉粒狀態。

裝飾步驟

1. **填入咖椰醬：**在蛋糕中間挖1小孔，填入咖椰醬。
2. **使用開縫星齒花嘴：**拿1個擠花袋，放入開縫法式花嘴。將咖椰奶油糖霜放入擠花袋，往下推實，紮緊袋口，在蛋糕表面擠上花紋。
3. **放上咖椰吐司：**將吐司切小塊，每1塊小吐司塗上1層薄咖椰醬。將咖椰吐司放在每1個蛋糕上作裝飾。

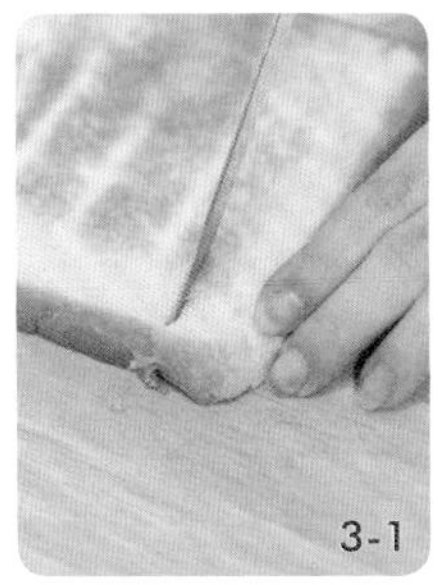
3-1

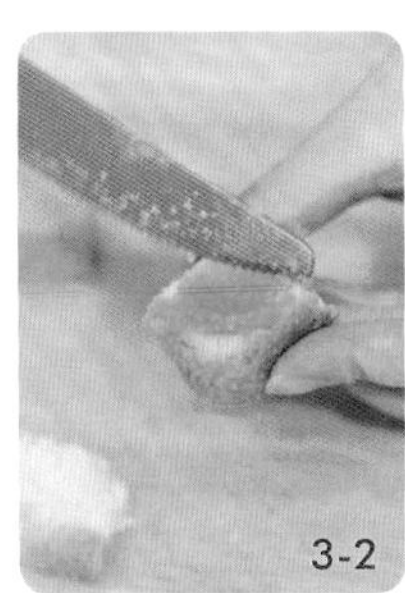
3-2

Kaya Cupcakes

INGREDIENTS

For Cupcake
55g unsalted butter
(at room temperature)
50g caster sugar
1 large egg
100g self-raising flour
40ml coconut milk
25g vegetable oil
2 tablespoons kaya paste

For Kaya Buttercream
100g unsalted butter
(at room temperature)
180g icing sugar
2-3 teaspoons coconut milk
2 teaspoons kaya paste

For Decoration
French Star Piping Tip
1 piping bag
6 pieces of mini toasted bread
extra kaya paste
(for cake filling and spreading on toasted bread)

CUPCAKE STEPS

1. Preheat the oven to 170 degrees Celsius.
2. Beat the butter and sugar until smooth.
3. Add in the egg and mix well.
4. Sift the self-raising flour. Add in the flour little by little to the batter.
5. Add in the coconut milk little by little. Add in the vegetable oil. Stir in the kaya paste.
6. Pour into cupcake cups, 2/3 full. Bake for 20-30 minutes or until golden brown.
7. Transfer to cooling rack. Cool down completely before frosting.

KAYA BUTTERCREAM STEPS

1. Beat the unsalted butter until soft. Sift the icing sugar and add in 6-7 times.
2. Add in the milk one teaspoon at a time.
3. Stir in the Kaya paste.
4. Beat until smooth and creamy.

DECORATION STEPS

1. Cut a hole in each cupcake. Fill each centre with kaya paste.
2. Place the French Star Piping Tip into a piping bag. Place the Kaya Buttercream into the piping bag and pipe onto each cupcake.
3. Spread some kaya paste on each mini toast. Decorate with Mini Kaya Toast on each cupcake.

班蘭蛋糕

班蘭蛋糕是新加坡的特產，也是最受歡迎的伴手禮！在機場已經有很多店鋪可以買到。班蘭葉有一種獨特的天然香氣，能給食物增添清新香甜的味道。這種植物在新加坡非常普遍，當地人會用於雞飯、果醬、麵包、餅乾裡，用途很廣！現在我把班蘭味道放進cupcake，這樣在家也可以隨時吃到星洲美味啦！

材料

蛋糕

- □ 無鹽奶油 45克　□ 白砂糖 70克　□ 雞蛋 1顆　□ 班蘭香精 1/2茶匙
- □ 自發粉 90克　□ 椰奶 40毫升　□ 植物油 20克　□ 綠色食用色素 適量（可選擇不加）

班蘭法式蛋白奶油霜

- □ 班蘭香精 1/2茶匙　□ 綠色食用色素 1～2滴
- □ 法式蛋白奶油霜 1份（作法詳見23頁）

裝飾

- □ 擠花袋 1個　□ 糖花適量　□ 小吸管 6支　□ 開縫星齒花嘴

製作蛋糕

1 **準備：**將烤箱預熱至170℃。奶油在室溫中回軟。

2 **打發奶油：**加入白砂糖，和奶油一起打發，直到顏色變淺，呈軟滑狀。

3 **加入雞蛋：**加入打散的蛋液，均勻攪拌，直到和奶油混合。

4 **篩入粉類：**自發粉過篩，然後分3次加入。用矽膠刮刀以切拌的方式快速拌勻。

5 **加入其他材料：**慢慢加入椰奶。再加入植物油和班蘭香精。

6 **入烤箱烘焙：**將麵糊倒入蛋糕杯，約2/3滿。用170℃烤20～30分鐘或烤至蛋糕面呈金黃色。

7 **冷卻：**完成後，轉移到散熱架上。等蛋糕完全冷卻後才裝飾。

製作班蘭法式蛋白奶油霜

1 **加入班蘭香精：**將班蘭香精加入預先準備好的法式蛋白奶油霜中。

2 **加入色素：**加綠色食用色素。用抹刀將色素與奶油霜混合均勻。

裝飾步驟

1 **使用開縫星齒花嘴：**拿1個擠花袋，放入開縫星齒花嘴。將班蘭法式蛋白奶油霜放入擠花袋，往下推實，紮緊袋口，在蛋糕表面擠上花紋。

2 **放上裝飾物：**在每個蛋糕上放上1支小吸管和幾朵糖花作裝飾。

Pandan Cupcakes

INGREDIENTS

For Cupcake

45g unsalted butter
(at room temperature)
70g caster sugar
1 large egg
90g self-raising flour
40ml coconut milk
(at room temperature)
20g vegetable oil
1/2 teaspoon pandan essence
green food coloring (optional)

For Pandan French Meringue Buttercream

1 portion of French Meringue Buttercream
(please refer to page 23)
1/2 teaspoon pandan essence
1-2 drops of green food coloring

For Decoration

Open Star Piping Tip
1 piping bag
6 drinking straws
sugar flowers

CUPCAKE STEPS

1.Preheat the oven to 170 degrees Celsius.
2.Beat the butter and sugar until smooth.
3.Add in the egg and mix well.
4.Sift the self-raising flour. Add in the flour little by l ittle to the batter.
5.Add in the coconut milk little by little. Add in the vegetable oil and the pandan essence.
6.Pour into cupcake cups, 2/3 full. Bake for 20-30 minutes or until golden brown.
7.Transfer to cooling rack. Cool down completely before frosting.

PANDAN FRENCH MERINGUE BUTTERCREAM STEPS

1.Add the pandan essence into 1 portion o f French Meringue Buttercream.
2.Add in the green food coloring. Mix until smooth and creamy.

DECORATION STEPS

1.Place the Open Star Piping Tip into a piping bag. Place the Pandan Buttercream into the piping bag and pipe onto each cupcake.
2.Decorate with Straw and sugar flower.

英國
UNITED KINGDOM

說到英國飲食，相信大家最熟悉的就是英式早餐和High Tea。
不過很多人誤以為High Tea是下午茶的意思，
其實傳統英國人所說的「High Tea」或「Meat Tea」指的是晚餐。
至於下午茶，應該叫做「Low Tea」，而且還分為三類。
第一類「Cream Tea」只有茶、鬆餅和果醬；
第二類「Light Tea」會再加少許甜點；
而第三類「Full Tea」通常還會加三明治、蛋糕、餅乾等等。
英國菜的烹調方法通常比較簡單，但或許這正是他們的特色。
英國人對於餐桌禮儀相當講究，連用的餐具都大有名堂，
這也是英國成為優雅、高貴的代名詞的原因吧！

香橙巧克力蛋糕

香橙加巧克力可以說是最佳搭配！而英國人最愛的甜食Jaffa cakes（佳發蛋糕）就是香橙巧克力加蛋糕。為了做一個cupcake版本，我選擇用巧克力蛋糕做底，加入新鮮的橙皮、香橙果醬和橙汁來製造濃烈的香橙味道，頂部則用了巧克力醬！最後，將新鮮香橙塊烤乾，再沾上巧克力。這樣就可以確保香橙味和巧克力味一樣濃烈！

材料

蛋糕

- 橙汁 1/2個
- 橙皮乾 15克
- 基本巧克力蛋糕麵糊 1份（作法詳見20頁）

香橙奶油糖霜

- 橙皮 1/2個
- 香橙香精 1/2茶匙
- 橙紅色食用色素 1～2滴
- 基本奶油糖霜 1份（作法詳見21頁）

裝飾

- 新鮮橙片 6片
- 擠花袋 1個
- 開縫法式花嘴
- 巧克力 適量
- 香橙果醬

製作蛋糕

1 **混合橙汁和蛋糕糊**：將橙汁倒入預先備好的基本巧克力蛋糕麵糊裡。

2 **加入橙皮乾**：加入橙皮乾，拌勻。

3 **入烤箱烘焙**：將麵糊慢慢倒入蛋糕杯約2/3滿。用170℃烤20～30分鐘或烤至蛋糕表面呈金黃色。

4 **冷卻**：完成後，轉移到散熱架上。等蛋糕完全冷卻後才裝飾。

製作香橙奶油糖霜

1 **混合橙皮和奶油糖霜**：將橙皮加入預先準備好的基本奶油糖霜裡。

2 **加入香橙香精**：將香橙香精倒入奶油糖霜裡。

3 **加入橙紅色色素**：加入橙紅色食用色素。

4 **打發**：打發至鬆軟、無粉粒狀態。

裝飾步驟

1 **填入果醬**：在蛋糕中間挖1個小孔，填入香橙果醬。

2 **使用開縫法式花嘴**：使用開縫法式花嘴，將香橙奶油糖霜擠上蛋糕表面。

3 **放上裝飾品**：放1片新鮮橙片，或1片香橙巧克力片作裝飾。

1-1

1-2

小技巧／製作香橙巧克力片

1.切橙片→將橙切成大約0.3公分薄片，然後把橙片平鋪在烘焙紙上。
2.烤乾→放入烤箱低溫度80℃烤1.5～2小時，直到水分乾涸。
3.巧克力隔水加熱→將黑巧克力放入碗裡，隔水加熱。
4.沾上巧克力醬→將橙片沾上一些巧克力醬，放回烘焙紙，於室溫擺放約1小時，即可使用。

1

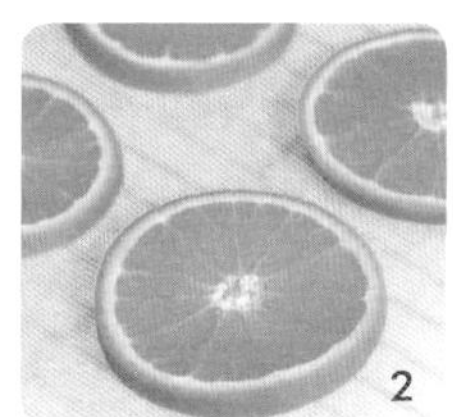
2

4-1

4-2

Jaffa Cake Cupcakes

INGREDIENTS

For Cupcake

1 portion of Basic Chocolate Cupcake Batter
(please refer to page 20)
1/2 orange juice
15g dried orange peel

For Orange Buttercream

1 portion of Basic Buttercream
(please refer to page 21)
1/2 orange zest
1/2 tsp orange essence
1-2 drops of orange food coloring

For Decoration

French Star Piping Tip
1 piping bag
orange jam (instant sugar free orange marmalade)
6 pieces of orange slices chocolate

CUPCAKE STEPS

1.Add 1/2 orange juice onto 1 portion of Basic Chocolate Cupcake Batter.
2.Fold in orange juice.
3.Pour into cupcake cups, 2/3 full. Bake for 20-30 minutes or until golden brown.
4.Transfer to cooling rack. Cool down completely before frosting.

BUTTERCREAM STEPS

1.Add the orange zest into 1 portion of Basic Buttercream.
2.Add in the orange essence.
3.Add in the orange food coloring.
4.Beat until smooth and creamy.

DECORATION STEPS

1.Cut a hole in each cupcake. Fill it with orange jam.
2.Place the French Star Piping Tip into a piping bag. Place the Orange Buttercream into the piping bag and pipe onto each cupcake.
3.Decorate with a piece of fresh orange slice or chocolate-coat dried orange slices.

TIPS : HOW TO MAKE CHOCOLATE-COAT DRIED ORANGE SLICES?

1.Cut orange into 3mm thin slices. Put orange slices on baking tray lined with parchment paper.
2.Bake orange slices in oven at low temp 80°C for 1.5 to 2 hours until water dried out.
3.Melt some dark chocolate in a bowl.
4.Dipped the dried orange slices into melt chocolate and put on parchment paper, let set in room temperature for one hour.

檸檬派蛋糕

檸檬派是我媽媽的最愛。記憶中這道甜品有很多奶油和很香的檸檬味。為了保留這份甜蜜回憶，我將它做成杯子蛋糕。很多人爭論檸檬派的起源，有人說源自英國，有人說是法國，也有人說是美國，但還沒有定論。不過，網路上有一個比較接近的說法，就是早在1940年，英國已經有一個叫做「Chester Pudding」的甜點跟檸檬派很相似。

材料

脆餅底

- 無鹽奶油 40克
- 消化餅乾碎 80克

蛋 糕

- 檸檬汁 2湯匙
- 檸檬皮 1/2個
- 基本香草奶油蛋糕麵糊 1份（作法詳見19頁）

檸檬酪醬

- 雞蛋1 顆
- 檸檬汁 20毫升
- 檸檬皮 1/4湯匙
- 白砂糖 30克
- 無鹽奶油 15克

裝 飾

- 蛋白霜 1份（作法詳見22頁）
- 開縫星齒花嘴
- 新鮮檸檬切片或檸檬形狀糖果
- 擠花袋 1個

製作蛋糕

1 **製作脆餅底：**奶油融化後，加入餅乾碎裡。將餅乾碎放入蛋糕杯的底部，壓平至0.4～0.5公分厚。

2 **製作檸檬麵糊：**將檸檬汁和檸檬皮拌勻，加入基本香草奶油蛋糕麵糊裡。

3 **入烤箱烘焙：**將麵糊慢慢倒入蛋糕杯，約2/3滿。用170℃烤20～30分鐘或烤至蛋糕表面呈金黃色。

4 **冷卻：**完成後移到散熱架上。等蛋糕完全冷卻後才裝飾。

1-1 1-2 1-3 2 3 4

製作檸檬酪醬

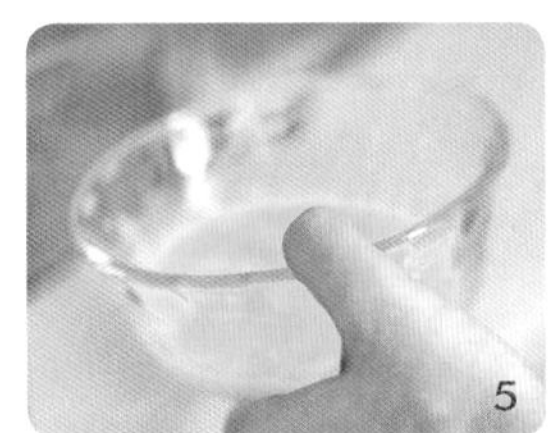

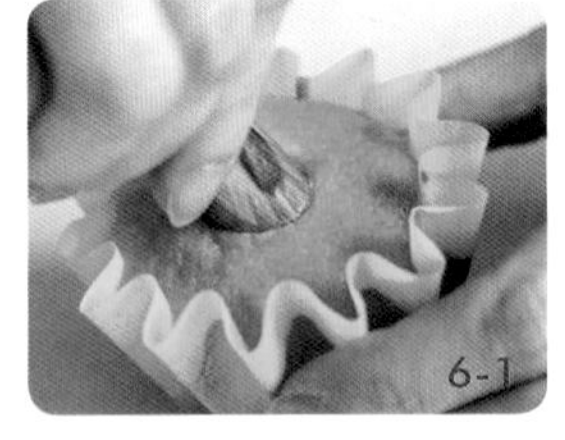

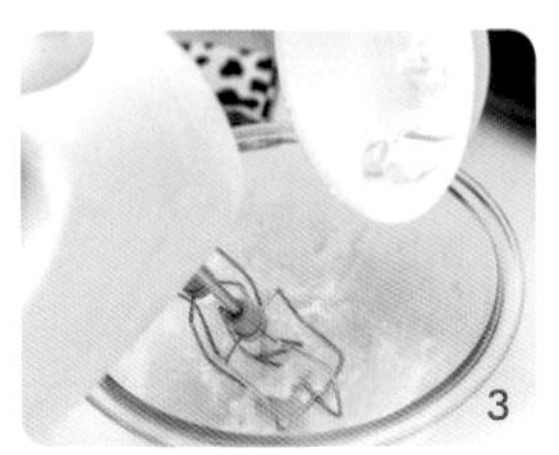

1. **隔水加熱**：將雞蛋、糖、檸檬汁放在碗中拌勻，隔水加熱直到混合。離火，持續攪拌約10分鐘，直到混合物變得濃稠。
2. **過濾**：將混合物過篩，以去除團塊。
3. **加入奶油**：將奶油切小塊加入混合物中，攪拌至完全混合。
4. **加入檸檬皮**：將檸檬皮倒入混合物中，攪拌均勻。
5. **入冰箱冷藏**：待它冷卻後，放入冰箱冷藏1天備用。
6. **填餡**：在蛋糕中間挖1小孔，填入檸檬酪醬。*也可到超市買即食檸檬蛋黃醬。

裝 飾 步 驟

1. **使用開縫星齒花嘴**：拿1個擠花袋，放入開縫星齒花嘴。將蛋白霜放入擠花袋，往下推實，紮緊袋口，在蛋糕表面擠上花紋。
2. **將蛋白霜表面燒至金黃色**：使用烘焙用噴火槍，將蛋白霜表面燒至金黃色。
3. **放上檸檬**：最後，放1片新鮮檸檬片當裝飾。

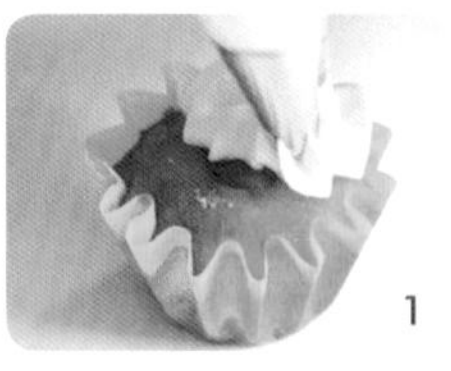

Lemon Meringue Pie Cupcakes

INGREDIENTS

For Bottom Layer
40g unsalted butter (melted)
80g digestive biscuits (crumbs)

For Cupcake
1 portion of Basic Vanilla Cupcake Batter
(please refer to page 19)
2 tablespoon lemon juice
zest of 1/2 lemon

For Lemon Filling
1 large egg
20ml freshly squeezed lemon juice
1/4 tablespoon finely shredded lemon zest
30g granulated white sugar
15g unsalted butter (at room temperature)

For Decoration
Open Star Piping Tip
1 piping bag
1 portion of Meringue Frosting
(please refer to page 22)
Fresh lemon slices or lemon jelly candies

CUPCAKE STEPS

1. For the Bottom:
   Melt the butter, add in digestive biscuits to make into crumbs. Place the mixture and press into flat layer 4-5mm on bottom of each cupcake.
2. For the Cupcake:
   Add the lemon juice and lemon zest to 1 portion of Basic Vanilla Cupcake Batter.
3. Pour into cupcake cups, 2/3 full. Bake for 30-40 minutes or until golden brown.
4. Transfer to cooling rack. Cool down completely before frosting.

FILLING STEPS

1. Whisk the eggs, sugar and lemon juice in a mixing bowl over a saucepan of simmering water, until blended. Stir constantly until mixture becomes thickened (around 10 minutes).
2. Remove from heat, pour through fine strainer to remove lumps.
3. Cut the butter in small pieces and whisk it into the mixture until co-operated.
4. Add in the lemon zest.
5. Let it cool and refrigerate overnight.
6. Cut a hole in each cupcake. Fill each centre with lemon filling.

DECORATION STEPS

1. Place the Open Star Piping Tip into a piping bag. Place the Meringue Frosting into the piping bag and pipe onto each cupcake.
2. Use a flaming torch to slightly burn meringue to light brown color.
3. Decorate with a slice of fresh lemon or lemon jelly candy.

麵包布丁蛋糕

麵包布丁是英國很多傳統家庭都喜歡的甜點！原因是簡單易做，只需要用剩餘的麵包混合其他材料，不僅操作容易，還可以避免食物的浪費。我自己喜歡用葡萄麵包，加入些許雞蛋、奶油、牛奶和肉桂粉來做麵包布丁。於是，我將這個簡單的麵包布丁食譜融入cupcake，製造一個充滿英式風味的麵包布丁蛋糕。

材料

蛋糕

- 葡萄乾 1/4杯
- 基本香草奶油蛋糕麵糊 1份（作法詳見19頁）

麵包布丁

- 葡萄乾麵包 3片
- 肉桂粉 1茶匙
- 雞蛋1顆
- 牛奶 40毫升
- 鮮奶油 10毫升

製作步驟

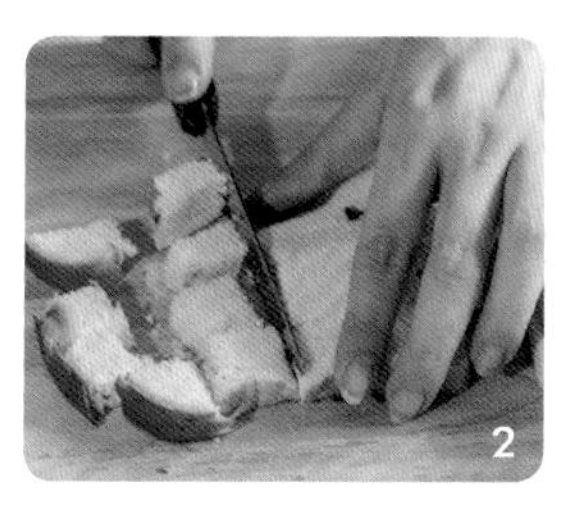
2

3-3

3-1

3-4

3-2

4

1 **放入葡萄乾：**將葡萄乾放入預先準備好的基本香草奶油蛋糕麵糊裡。將麵糊慢慢倒入蛋糕杯，約2/3滿。

2 **麵包切片：**將葡萄乾麵包切成小方片。

3 **混合：**將肉桂粉、雞蛋、牛奶和鮮奶油倒入1個大碗內，混合均勻。然後將葡萄乾麵包片放入其中。

4 **放在蛋糕頂部：**將浸透的小方片麵包放在蛋糕的頂部。

5 **入烤箱烘焙：**在蛋糕麵糊上平鋪麵包布丁，入烤箱用170℃烤20～30分鐘或烤至蛋糕表面呈金黃色。

Bread Pudding Cupcakes

INGREDIENTS

For Cupcake

1 portion of Basic Vanilla Cupcake Batter
(please refer to page 19)
1/4 cup raisins

For Bread Pudding Topping

3 pieces of raisin bread
1 teaspoon grounded cinnamon powder
1 egg
40ml milk
10ml whipping cream

CUPCAKE STEPS

1. Stir in the raisins to 1 portion of Basic Vanilla Cupcake Batter. Pour into cupcake cups, 2/3 full.
2. Cut the bread into small square pieces.
3. Mix the grounded cinnamon, egg, milk and cream in a cup. Dip the bread pieces into the cream mixture.
4. Place a few pieces on top of the cupcakes.
5. Bake together with the cupcakes for 20-30 minutes or until golden brown.

格雷伯爵茶蛋糕

關於格雷伯爵茶的起源，傳說是因為格雷伯爵依自己的喜好將不同的茶葉混合，並加入具安定效用的香檸檬油，後來傳入民間，廣受大眾歡迎。英國人很喜歡享受下午茶，他們常常會一邊喝茶，一邊吃甜點。所以我從中得到靈感，將格雷伯爵茶加入蛋糕裡面，再將蛋糕放進茶杯裡面，做出一杯可以吃的「茶」，一次滿足兩個要求！

材料

蛋糕

- 格雷伯爵茶 20毫升（先用熱水泡）
- 格雷伯爵茶葉 1茶匙
- 基本香草奶油蛋糕麵糊 1份（作法詳見19頁）

裝飾

- 擠花袋 1個
- 鮮奶油 250毫升
- 格雷伯爵茶葉
- 開縫星齒花嘴

製作蛋糕

1. **加入格雷伯爵茶：**將格雷伯爵茶用熱水沖泡，泡好後漸漸倒入預先備好的基本香草奶油蛋糕麵糊中。
2. **加入茶葉：**將格雷伯爵茶葉倒入蛋糕糊，拌勻。
3. **入烤箱烘焙：**將麵糊慢慢倒入蛋糕杯，約2/3滿。用170℃烤20～30分鐘或烤至蛋糕表面呈金黃色。
4. **冷卻：**完成後，轉移到散熱架上。等蛋糕完全冷卻後才裝飾。

1

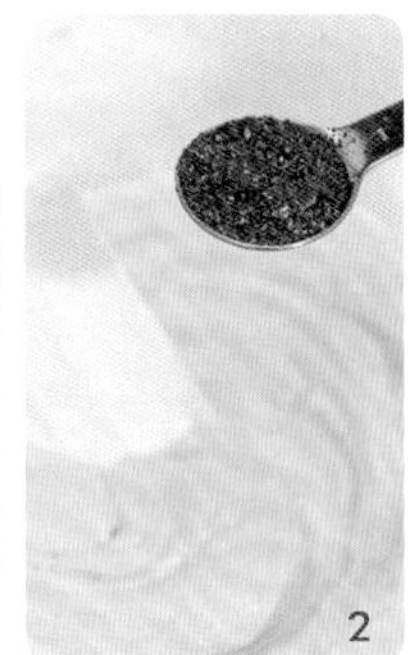
2

裝飾步驟

1. **打發鮮奶油：**將鮮奶油打發至硬性發泡。
2. **使用開縫星齒花嘴：**拿1個擠花袋，放入開縫星齒花嘴。將鮮奶油放入擠花袋，往下推實，紮緊袋口，在蛋糕表面擠上花紋。
3. **撒上格雷伯爵茶葉：**在蛋糕表面上再撒一些格雷伯爵茶葉作裝飾。
4. **放入茶杯模具：**最後，把蛋糕放入特別的茶杯模具內，美化外觀。

Earl Grey Tea Cupcakes

INGREDIENTS

For Cupcake

1 portion of Basic Vanilla Cupcake Batter (please refer to page 19)
20ml Earl Grey tea (soaked in hot water)
1 teaspoon Earl Grey tea leaves

For Decoration

Open Star Piping Tip
1 piping bag
250ml whipped cream
Earl Grey tea leaves (sprinkle on top)

CUPCAKE STEPS

1. Add in the Earl Grey tea (water) little by little into 1 portion of Basic Vanilla Cupcake Batter.
2. Stir in the Earl Grey tea leaves (dried).
3. Pour into cupcake cups, 2/3 full. Bake for 20-30 minutes or until golden brown.
4. Transfer to cooling rack. Cool down completely before frosting.

DECORATION STEPS

1. Whip the cream until peak.
2. Place the Open Star Piping Tip into a piping bag. Place the whipped cream into the piping bag and pipe onto each cupcake.
3. Sprinkled with dried Earl Grey tea leaves.
4. Served on teacup.

法國
FRANCE

法國是浪漫之都。除了人浪漫，就連食物的文化都一樣浪漫。
因為浪漫是需要時間和心思來營造的。
吃一頓法國餐要幾個小時，品嘗紅酒、白酒要時間，
做法國甜點，比如馬卡龍也需要很長時間……
可以說是慢工出細活！法國人很有心思，
吃東西除了食物要好吃，環境、情調一樣很重要。
烹調食物的時候，也會用很多不同的程序來達到這個目的，
所以法式甜點特別精巧繁複。

薰衣草蛋糕

我最喜歡的顏色是紫色，而薰衣草就是這種顏色。薰衣草顏色美，香味也很獨特。很多人會用薰衣草做香精油，因為它有舒緩的功效。我收過最特別的薰衣草禮物，是一位嫁給法國人的好朋友送的薰衣草香包。她跟我說這些香包是她老公家鄉的特產，只要將香包放在衣櫃裡，整個空間就會瀰漫著沁人心田的香氣。這款薰衣草cupcake，選用了薰衣草乾和薰衣草食用香精做成。希望大家在吃完之後，會有一種放鬆心情的感覺。

材料

蛋糕

- 基本香草奶油蛋糕麵糊 1份（作法詳見19頁）
- 薰衣草乾葉碎 2茶匙
- 薰衣草水 10毫升（先用熱水泡薰衣草）
- 紫色食用色素 1～2滴（可選擇不加）
- 薰衣草香精 1/2茶匙（可選擇不加）

薰衣草奶油糖霜

- 基本奶油糖霜 1份（作法詳見21頁）
- 薰衣草香精 1/2茶匙
- 紫色食用色素 1滴

裝飾

- 薰衣草乾葉碎
- 迷你馬卡龍
- 圓口花嘴
- 擠花袋 1個
- 巴黎鐵塔紙牌裝飾

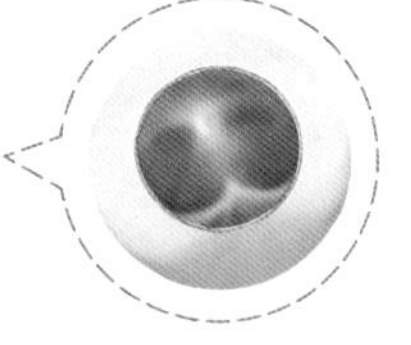

製作蛋糕

1. **混合薰衣草水和蛋糕糊：**將薰衣草水放入預先準備好的基本香草奶油蛋糕麵糊裡。
2. **加入其他材料：**加入薰衣草乾葉碎、薰衣草香精和紫色食用色素。
3. **入烤箱烘焙：**將麵糊慢慢倒入蛋糕杯，約2/3滿。用170℃烤20～30分鐘或烤至蛋糕表面呈金黃色。
4. **冷卻：**完成後，轉移到散熱架上。等蛋糕完全冷卻後才裝飾。

1

2

3

4

製作薰衣草奶油糖霜

1

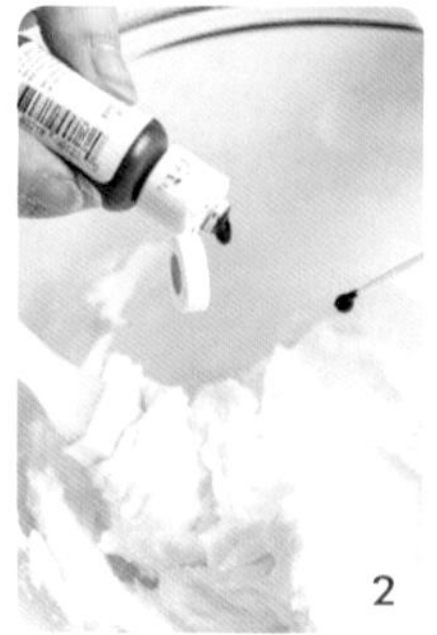
2

3

1. **加入薰衣草香精：**將薰衣草香精加入預先準備好的基本奶油糖霜裡。
2. **加入色素：**加入紫色食用色素，攪拌至奶油糖霜呈淡紫色。
3. **打發：**將糖霜打發至鬆軟、無粉粒狀態。

裝飾步驟

1 **使用圓口花嘴**：拿1個擠花袋，放入圓口花嘴。將薰衣草奶油糖霜放入擠花袋，往下推實，紮緊袋口，在蛋糕表面擠上1層糖霜，再用抹刀抹平。

2 **撒上薰衣草乾葉碎**：在糖霜上撒上薰衣草乾葉碎。

3 **放上馬卡龍**：在每個蛋糕上面放1個迷你馬卡龍甜點作裝飾。

4 **插上裝飾牌**：最後，在蛋糕上插1個巴黎鐵塔紙牌作裝飾。

1-1

1-2

2

3

4

Lavender Cupcakes

INGREDIENTS

For Cupcake

1 portion of Basic Vanilla Cupcake Batter (please refer to page 19)
10ml lavender water (dried lavender dissolved in hot water)
2 teaspoon dried lavender
l teaspoon lavender essence (optional)
1-2 drops of purple food coloring (optional)

For Lavender Buttercream

1 portion of Basic Buttercream (please refer to page 21)
1/2 teaspoon lavender essence
1 drop of purple food coloring

For Decoration

Round Piping Tip
1 piping bag
Mini Macaroon
dried lavender
Eiffel Towel paper topper

CUPCAKE STEPS

1.Add the lavender water little by little into 1 portion of Vanilla Cupcakes.
2.Stir in dried lavender. Addin the lavender essence (optional). Add in a drop of purple food coloring (optional).
3.Pour into cupcake cups, 2/3 full. Bake for 20-30 minutes or until golden brown.
4.Transfer to cooling rack. Cool down completely before frosting.

LAVENDER BUTTERCREAM STEPS

1.Add lavender essence to 1 portion of Basic Buttercream.
2.Add in the purple food coloring.
3.Mix until well blended.

DECORATION STEPS

1.Place the Round Piping Tip into a piping bag. Place the Lavender Buttercream into the piping bag and pipe onto each cupcake. Smooth out with hand spatula.
2.Sprinkle dried lavender on top of each cupcake.
3.Topped with Macaroon.
4.Place an Effiel Tower paper topper as decoration.

洋梨夏洛蒂蛋糕

Charlotte是法國一種甜點的名稱。通常都是用一般奶油蛋糕，加入卡士達或果醬，然後放上新鮮的生果。我喜歡洋梨夏洛蒂，因為西洋梨的口感很鬆軟，與卡士達是最佳配搭。這個甜點另一個特別之處就是它的蛋糕由手指餅圍繞著。

材料

蛋糕

- 基本香草奶油麵糊 1份（作法詳見19頁）

裝飾

- 即食卡士達奶黃醬 適量
- 鮮奶油 100毫升
- 手指餅乾 1包
- 西洋梨 1個（切成片）
- 絲帶半吋

製作步驟

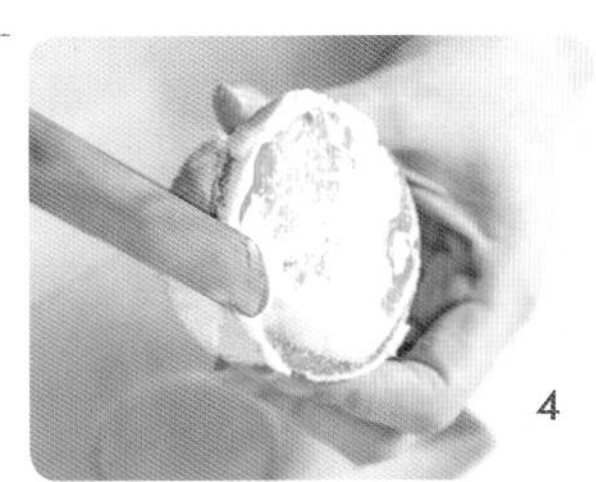
4

1 **入烤箱烘焙：**將基本香草奶油麵糊慢慢倒入蛋糕杯，約2/3滿。用170℃烤20～30分鐘或烤至蛋糕表面呈金黃色。

2 **填入卡士達奶黃醬：**在每個蛋糕中間挖1個小孔，填入卡士達奶黃醬。

3 **打發鮮奶油：**將鮮奶油打發至硬性發泡。

4 **塗抹鮮奶油：**將鮮奶油抹上蛋糕表面及側面。

5 **用手指餅乾圍住蛋糕：**用手指餅乾圍住整個蛋糕。用1條絲帶蝴蝶結圍繞手指餅作裝飾。

6 **放上西洋梨片：**在中間放一些西洋梨片。

7 **製造鏡面效果（可選擇不加）：**最後，在蛋糕表面刷上果膠即成。

5

小技巧／自製果膠

材料：**糖1/2杯 果汁1杯 玉米粉2湯匙**

1.糖和1/2杯果汁放入小鍋中，以中火煮熱。

2.將玉米粉倒入剩下的1/2杯果汁中，加入小鍋中一起煮。煮至果膠變濃。

★注意：你所選的果汁會影響到果膠的顏色和味道。

1

2-1

2-2

2-3

Pear Charlotte Cupcakes

INGREDIENTS

For Cupcake

1 portion of Basic Vanilla Cupcake Batter (please refer to page 19)

For Decoration

100ml whipped cream
devon custard
1 packet of lady finger biscuits
1 pear (cut into slices)
1/2 inches width ribbon tape

STEPS

1. Pour Basic Vanilla Cupcake Batter into cupcake cups, 2/3 full. Bake for 20-30 minutes or until golden brown.
2. Cut a hole in each cupcake. Fill each with devon custard.
3. Whip the cream until peak.
4. Spread the whipped cream on the sides of each cupcake.
5. Cut the lady fingers in half and place around each cupcake. Wrap a ribbon around the lady fingers.
6. Place the pear slices in the middle.
7. Topped with a sugar glaze on top of the pears (optional).

TIPS : How To Make Sugar Fruit Glaze?

Ingredients

1/2 cup sugar
1 cup fruit juice
2 tablespoons cornstarch

Steps

1. Put the sugar and 1/2 cup of fruit juice in a saucepan and bring to boil.
2. Dissolve the cornstarch into the remaining 1/2 cup of fruit juice and pour it into the saucepan mixture. Cook until solution thickens.

* The type of fruit juice you choose will affect the flavor and the color of the glaze.

焦糖布丁蛋糕

傳統的焦糖布丁比較creamy，很多人會覺得太甜，而且有些人喜歡蛋味較濃的味道，所以不少布丁慢慢演變得像燉蛋。焦糖布丁的重點當然就在布丁的表面，燒出來的焦糖，不僅外表美觀，口感也很特別。我做的這個焦糖布丁cupcake保留了脆口的焦糖topping，同時蛋糕的質感與傳統布丁又有很大不同，相信一定能給你們帶來全新的體驗。

材料

蛋糕

- 基本香草奶油蛋糕麵糊 1份（作法詳見19頁）

裝飾

- 焦糖布丁（可自製或在超市買即用布丁醬）
- 白砂糖
- 新鮮草莓
- 烘焙用噴火槍

製作步驟

1. **填入布丁**：在每1個蛋糕中間挖1個大孔，填入焦糖布丁。
2. **撒上白砂糖**：撒一些白砂糖在布丁上面。
3. **使用噴火槍**：以噴火槍燒砂糖表面，直到呈現少許燒焦效果。
4. **放上新鮮草莓**：放上新鮮草莓作裝飾。

1-1

1-2

2

3

4

Crème Brulee Cupcakes

INGREDIENTS

For Cupcake

1 portion of Basic Vanilla Cupcake Batter (please refer to page 19)

For Decoration

custard pudding (can buy instant custard pudding from supermarket)
caster sugar
fresh strawberries
hand-gas-torch

STEPS

1.Cut a big hole in the surface of the cupcake. Frost the cake top with a flat layer of custard pudding.
2.Sprinkle some sugar on top.
3.Use hand gas-torch to caramelized the sugar.
4.Decorate with a fresh strawberry.

玫瑰蛋糕

《玫瑰人生》（La vie en rose）是法國最著名的香頌，玫瑰也是法國浪漫精神的象徵。玫瑰花可以用作裝飾，送給愛人表達情意，也可以食用。在法國到處都可以買到的玫瑰花瓣醬（Confit Petales de Rose），是將乾的玫瑰花瓣加入糖做成的。這種花醬在香港較難買到。我用玫瑰茶香粉替代，再用玫瑰奶油糖霜擠出花瓣作裝飾，做出這個玫瑰cupcake。相信會是一份很特別的情人節禮物！

材料

蛋糕

- 基本香草奶油蛋糕麵糊 1份（作法詳見19頁）
- 玫瑰茶 10毫升（先泡熱水）
- 粉紅色食用色素 1～2滴（可選擇不加）
- 玫瑰茶香粉 1茶匙

玫瑰奶油糖霜

- 基本奶油糖霜 1份（作法詳見21頁）
- 粉紅色食用色素 1～2滴
- 玫瑰茶香粉 2茶匙

裝飾

- 擠花袋 1個
- 開縫星齒花嘴

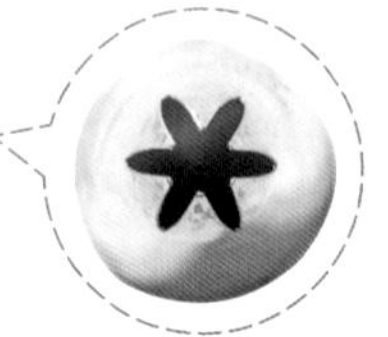

製作蛋糕

1｜**加入玫瑰茶：**將玫瑰茶溶液加入預先準備好的基本香草奶油蛋糕麵糊裡。

2｜**加入玫瑰茶香粉：**加入玫瑰茶香粉。

3｜**入烤箱烘焙：**將麵糊倒入蛋糕杯，約2/3滿。用170℃烤20～30分鐘或烤至蛋糕面呈金黃色。

4｜**冷卻：**完成後，轉移到散熱架上。等蛋糕完全冷卻後才裝飾。

製作玫瑰奶油糖霜

1｜**加入玫瑰茶香粉：**將玫瑰茶香粉加入預先準備好的基本奶油糖霜裡面。

2｜**加入粉紅色色素：**加入粉紅色食用色素。

3｜**打發：**打發至鬆軟、無粉粒狀態。

裝飾步驟

1

2-1

2-2

1｜**使用開縫星齒花嘴：**拿1個擠花袋，放入開縫星齒花嘴。將玫瑰奶油糖霜放入擠花袋，往下推實，紮緊袋口，在蛋糕表面擠上花紋。

2｜**擠出玫瑰花紋：**由中心開始向外繞圈，呈玫瑰花紋狀，完成。

Rose Cupcakes

INGREDIENTS

For Cupcake

1 portion of Basic Vanilla Cupcake Batter
(please refer to page 19)
10ml rose tea
(dissolved in hot water)
1 teaspoon rose tea powder
1-2 drops of pink food coloring (optional)

For Rose Buttercream

1 portion of Basic Buttercream
(please refer to page 21)
2 teaspoon rose-tea powder
1-2 drops of pink food coloring

For Decoration

Open Star Piping Tip
1 piping bag

CUPCAKE STEPS

1.Add the rose tea into 1 portion of Basic Vanilla Cupcake Batter.
2.Stir in rose tea powder.
3.Pour into cupcake cups, 2/3 full.Bake for 20-30 minutes or until golden brown.
4.Transfer to cooling rack. Cool down completely before frosting.

ROSE BUTTERCREAM STEPS

1.Add the rose tea powder to 1 portion of Basic Buttercream.
2.Add in the pink food coloring.
3.Mix until well blended.

DECORATION STEPS

1.Place the Open Star Piping Tip into a piping bag. Place the Rose Buttercream into the piping bag and pipe onto each cupcake.
2.Pipe from centre out and around to get a simple swirl rose effect.

義大利
ITALY

義大利是一個文明古國，

歷史可追溯至大約20萬年前的舊石器時代。

許多地區都有重要的考古遺址，

有很多值得觀賞的建築物、藝術品。

義大利的美食種類很多，有披薩、義大利麵、香腸、

海鮮、起司等等，光是義大利麵和起司就已經超過300種！

不同城市所做的義大利菜口味都不一樣。

不過讓我印象深刻的還是義大利的人情味，

很多去過義大利的朋友都不懂義大利文，

但總會被義大利人的熱情所打動。

如果你問一個義大利人哪裡的義大利菜最好吃？

他一定會說是媽媽在家煮的菜。

提拉米蘇蛋糕

每間餐廳做的Tiramisu味道雖然差不多，但層次可能會有些許不同。因為有些喜歡加入很多奶油，有些會有蛋糕部分，有些會有起司部分。不論層次多與寡，濃郁的咖啡味道絕對少不了。我做的這款Tiramisu味道的cupcake，保留了傳統做法中的咖啡味和手指餅，裝飾上則以簡約為主。Tiramisu在義大利語中的意思是「把我拿起」，希望大家做完這個Tiramisu cupcake，都會很快拿起來享用！

材料

蛋糕

- 無鹽奶油 45克
- 白砂糖 70克
- 雞蛋 1顆
- 自發粉 90克
- 脫脂牛奶 30毫升
- 植物油 20克
- 咖啡香精 1/2茶匙（可選擇不加）
- 咖啡 20毫升（先溶於20毫升熱水）

裝飾

- 擠花袋 1個
- 鮮奶油或馬斯卡彭起司 250毫升
- 手指餅乾（可選）
- 無糖可可粉 適量
- 圓口花嘴

製作蛋糕

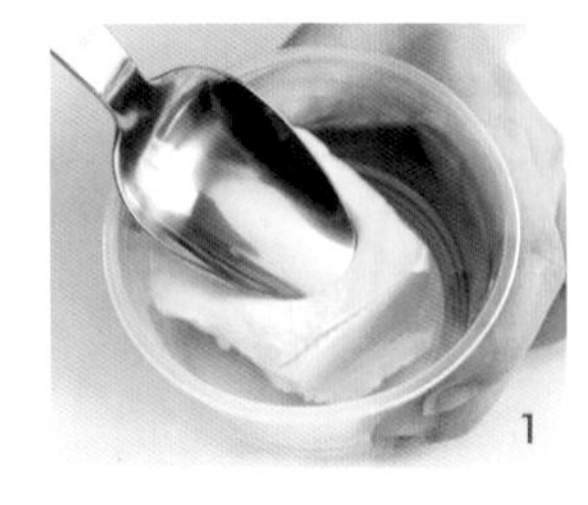
1

5-1

2

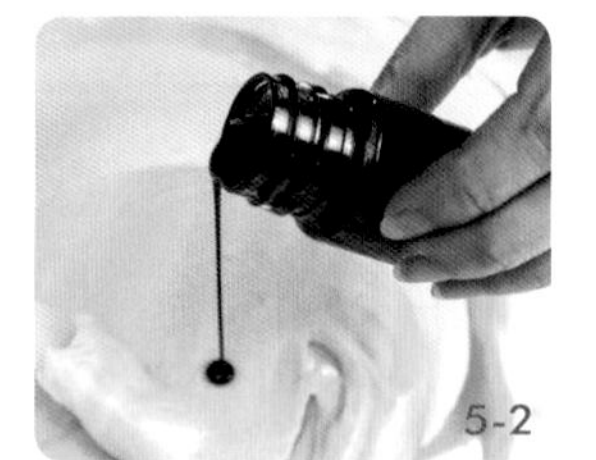
5-2

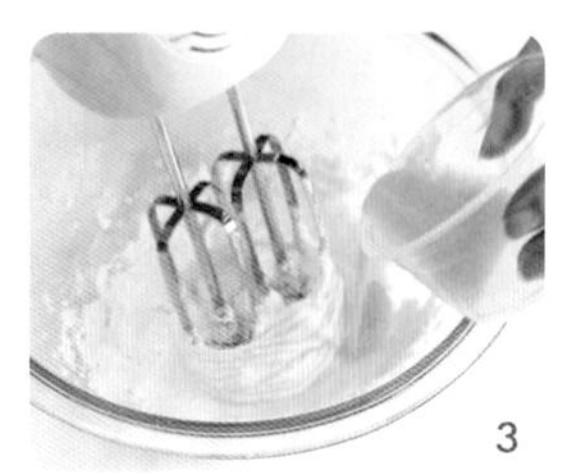
3

6

4

7

1. **準備**：將烤箱預熱至170℃。奶油在室溫中回軟。
2. **打發奶油**：加入白砂糖，和奶油一起打發，直到顏色變淺，呈軟滑狀。
3. **加入雞蛋**：加入打散的蛋液，混合攪拌，直到和奶油融合。
4. **篩入粉類**：將自發粉過篩，然後分3次加入。用切拌法快速拌勻。
5. **加入其他材料**：將脫脂牛奶跟咖啡液拌勻，漸漸加入麵糊中。再加入植物油和咖啡香精。
6. **入烤箱烘焙**：將麵糊慢慢倒入蛋糕杯，約2/3滿。用170℃烤20～30分鐘或烤至蛋糕表面呈金黃色。
7. **冷卻**：完成後移到散熱架上。等蛋糕完全冷卻後才裝飾。

裝飾步驟

1 **打發鮮奶油**：將鮮奶油打至硬性發泡。

2 **使用圓口花嘴**：拿1個擠花袋，放入圓口花嘴。將鮮奶油放入擠花袋，往下推實，紮緊袋口，在蛋糕表面擠上花紋。

3 **撒可可粉**：在蛋糕表面撒上無糖可可粉。

4 **放上手指餅**：最後，可放上1塊手指餅，再撒上可可粉和食用金箔作裝飾。

Tiramisu Cupcakes

INGREDIENTS

For Cupcake

45g unsalted butter
(at room temperature)
70g caster sugar
1 large egg
90g self-raising flour
30ml skimmed milk
20ml coffee (dissolved in 20ml hot water)
20g vegetable oil
1/2 teaspoon coffee essence (optional)

For Decoration

Round Piping Tip
1 piping bag
250 ml whipped cream or use instant mascorpone cheese
cocoa powder (sprinkle on top)
lady fingers (optional)

CUPCAKE STEPS

1.Preheat the oven to 170 degrees Celsius.
2.Beat the butter and sugar until smooth.
3.Add in the egg and mix well.
4.Sift the self-raising flour. Add in the flour little by little to the batter.
5.Mix the milk and coffee and add in the batter little by little. Add in the vegetable oil and coffee essence.
6.Pour into cupcake cups, 2/3 full. Bake for 20-30 minutes or until golden brown.
7.Transfer to cooling rack, and cool down completely before frosting.

DECORATION STEPS

1.Whip the cream (or Mascorpone cheese - can use immediately and pipe on cakes. Do not need to whip or beat to form cream).
2.Place the Round Piping Tip into a piping bag. Place the Whipped Cream or Mascorpone Cheese into the piping bag and pipe onto each cupcake.
3.Sprinkle cocoa powder on top.
4.Topped with lady fingers (optional).

聖誕蛋糕

義大利的聖誕蛋糕源自米蘭。義大利人會在聖誕節或者新年期間吃這種蛋糕。不過近年全世界，甚至連香港，都很容易在超市買到這種聖誕蛋糕。這種蛋糕的做法很簡單，就是在麵包中加入乾果。不過通常麵包比較乾，所以我做了一個蛋糕版本。大家在聖誕節的時候可以考慮做這款cupcake送給朋友，他們一定會很開心！

材料

蛋糕

- □基本香草奶油蛋糕麵糊 1份（作法詳見19頁）
- □乾果 1/2杯

製作步驟

1. **放入乾果**：將乾果放入預先準備好的基本香草奶油蛋糕麵糊裡。
2. **入烤箱烘焙**：將麵糊慢慢倒入蛋糕杯，約2/3滿。用170℃烤20～30分鐘或烤至蛋糕表面呈金黃色。
3. **冷卻**：完成後，轉移到散熱架上。

1

Panettone Cupcakes

INGREDIENTS

For Cupcake

1 portion of Basic Vanilla Cupcake Batter (please refer to page 19)
1/2 cup of dried fruits

STEPS

1. Add the dried fruits into 1 portion of Basic Vanilla Cupcake Batter.
2. Pour into cupcake cups, 2/3 full. Bake for 20-30 minutes or until golden brown.
3. Transfer to cooling rack.

蒙布朗蛋糕

Mont Blanc是一款經典的義式甜點，它的名字來自一座叫Mont Blanc（白朗峰）的雪山，因此這個蛋糕的外形也很像一座山。我把Mont Blanc變成cupcake，重點是要用栗子餡擠出一座小山，然後再撒上糖粉，就像頂著白雪的山頂。有個小祕訣就是要做成這個像山的栗子餡，一定要用多孔小圓口花嘴！如果還想錦上添花的話，可以加一些食用金箔在上面。

材料

蛋糕

- 基本香草奶油蛋糕麵糊 1份（作法詳見19頁）
- 鮮奶油 100毫升

裝飾

- 即食栗子泥 1/2罐
- 擠花袋 2個
- 糖粉 適量
- 圓口花嘴
- 多孔小圓口花嘴

製作蛋糕

3

1 **烘焙：**將預先準備好的基本香草奶油蛋糕麵糊倒入蛋糕杯，約2/3滿。用170℃烤20～30分鐘或烤至蛋糕表面呈金黃色。

2 **打發鮮奶油：**將鮮奶油打發至硬性發泡。

3 **填入鮮奶油：**在蛋糕中間挖1個小孔，用鮮奶油填滿小孔。

裝飾步驟

1 **使用圓口花嘴：**拿1個擠花袋，放入圓口花嘴。在蛋糕中央擠1個圓錐體。

2 **使用多孔小圓口花嘴：**再拿1個擠花袋，放入多孔小圓口花嘴。將栗子泥放入擠花袋，往下推實，紮緊袋口，先在蛋糕中央擠1個椎體，再繞圈式地擠上花紋。

3 **撒上糖粉：**最後，在蛋糕面上撒糖粉作裝飾。

1-1

1-2

2-1

2-2

3

Mont Blanc Cupcakes

INGREDIENTS

For Cupcake

1 portion of Basic Vanilla Cupcake Batter (please refer to page 19)

100ml whipped cream

For Decoration

Round Piping Tip

Multi Small Round Dots Piping Tip

2 piping bags

1/2 can instant Chestnut Purree

icing sugar (for dusting)

CUPCAKE STEPS

1. Prepare 1 portion of Basic Vanilla Cupcake Batter. Pour into cupcake cups, 2/3 full. Bake for 20-30 minutes or until golden brown.
2. Whip the cream until peak.
3. Cut a hole in each cupcake. Fill center with whipped cream.

DECORATION STEPS

1. Place the Round Piping Tip into a piping bag and start to pipe out a small cone in the centre of cupcake.
2. Place another Multi Small Round Dots Piping Tip into a piping bag. Place the Chestnut Purree into the piping bag and pipe around the centre cone to form a coil.
3. Dusted with icing sugar.

金莎巧克力蛋糕

1982年，義大利糖果廠商費列羅（Ferrero）推出金莎巧克力，立即大受歡迎，成為逢年過節送禮的首選！如果大家想特別一點，不妨試試做個金莎巧克力cupcake送給別人。這個cupcake裡面有1粒金莎巧克力，頂部也有1粒，這樣每一口都可以吃到金莎啦！喜歡的話還可以再加一些果仁碎！

材料

蛋糕

☐金莎巧克力 6粒 ☐基本巧克力蛋糕麵糊 1份（作法詳見20頁）

裝飾

☐金莎巧克力 6粒 ☐擠花袋 1個 ☐巧克力醬 1份（作法詳見25頁） ☐開縫星齒花嘴 ☐果仁碎（可選擇不加）

製作蛋糕

1 **倒入麵糊：**將少許基本巧克力蛋糕麵糊倒入蛋糕杯。

2 **放入金莎巧克力作餡：**將金莎巧克力放在蛋糕杯的中間。

3 **再倒入麵糊：**將剩餘的基本巧克力蛋糕麵糊倒入蛋糕杯，約2/3滿。

4 **烘烤並冷卻：**用170℃烤20～30分鐘或烤至蛋糕表面呈金黃色。完成後，轉移到散熱架上。等蛋糕完全冷卻後才裝飾。

裝飾步驟

1 **使用開縫星齒花嘴：**拿1個擠花袋，放入開縫星齒花嘴。將巧克力醬放入擠花袋，往下推實，紮緊袋口，在蛋糕表面擠上花紋。

2 **放上金莎巧克力作裝飾：**在蛋糕表面上放1粒金莎巧克力作裝飾。

3 **撒上果仁：**最後，可選擇喜歡的果仁碎撒在蛋糕面上。

Ferrero Rocher Cupcakes

INGREDIENTS

For Cupcake

1 portion of Basic Chocolate Cupcake Batter (please refer to page 20)
6 Ferrero Rocher Chocolate

For Decoration

Open Star Piping Tip
1 piping bag
1 portion of Chocolate Ganache (please refer to page 25)
6 Ferrero Rocher chocolate as decoration
nuts (optional)

CUPCAKE STEPS

1. Pour a little of the Basic Chocolate Cupcake Batter into cupcake cups.
2. Place a Ferrero Rocher chocolate in the middle of each cupcake cups.
3. Continue to pour the remaining cupcake batter into cupcake cups, 2/3 full, until cover up the Ferrero Rocher chocolate.
4. Bake for 20-30 minutes or until golden brown. Transfer to cooling rack. Cool down completely before frosting.

DECORATION STEPS

1. Place the Open Star Piping Tip into a piping bag. Place the Chocolate Ganache into the piping bag and pipe onto each cupcake.
2. Decorate with Ferroro Rocher chocolate on top.
3. Topped with chopped nuts (optional).

比利時

比利時是世界公認的巧克力之都，
最有名的巧克力生產商大部分都來自這裡。
在19世紀初，比利時的巧克力主要是供給上流社會，
而現在，比利時已經有超過2千家巧克力店，
巧克力變成普羅大眾隨時都可以買到的甜食。
不過，因為巧克力越來越受歡迎，
這幾年可可豆的價格也節節攀升，過不久，
不知道巧克力會不會再度成為奢侈的食品呢？
除了巧克力之外，比利時的鬆餅、貽貝、薯條和啤酒都很出名。
如果有機會到比利時，不妨好好品嘗一番！

比利時鬆餅蛋糕

在比利時的大街小巷都可以買到鬆餅，甚至有移動鬆餅車，可以說是當地的國民美食！為了製作有鬆餅特色的蛋糕，我選擇用冰淇淋甜筒杯來做材料。冰淇淋甜筒的好處，就是可以跟蛋糕一起放進烤箱裡烤。做出來的效果也很可愛，就好像一個真正的冰淇淋一樣！

材料

蛋糕

- □無鹽奶油 90克
- □白砂糖 140克
- □雞蛋 2顆
- □脫脂牛奶 80毫升
- □植物油 40克
- □香草精 1茶匙
- □蛋糕甜筒杯 6個（超市可買到）
- □自發粉 180克

裝飾

- □蛋白霜 1份（作法詳見22頁）
- □擠花袋1個
- □開縫星齒花嘴
- □巧克力淋面醬（作法詳見25頁）
- □彩色巧克力糖珠
- □鬆餅

製作蛋糕

1

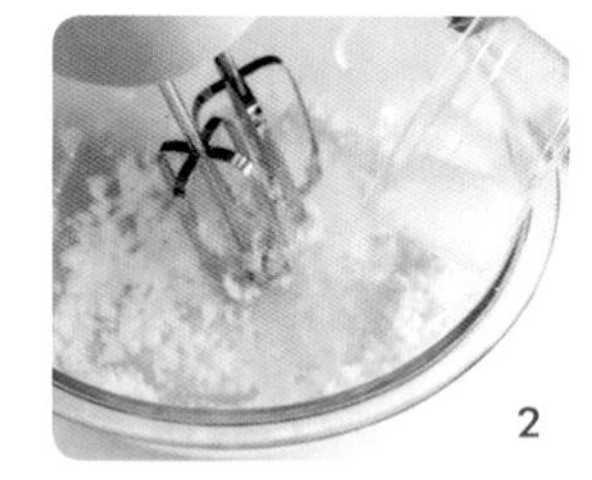
2

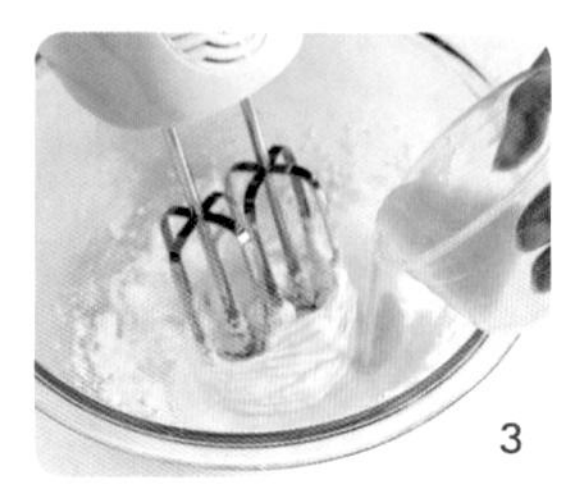
3

4

5

6

7

1. **準備：**將烤箱預熱至170℃。奶油在室溫中回軟。
2. **打發奶油：**加入白砂糖，和奶油一起打發，直到顏色變淺，呈軟滑狀。
3. **加入雞蛋：**加入打散的蛋液，混合攪拌，直到和奶油融合。
4. **篩入粉類：**將自發粉過篩，然後分3次加入。用矽膠刮刀以切拌的方式快速拌勻。
5. **加入其他材料：**加入脫脂牛奶、植物油、香草精，攪拌均勻。
6. **入烤箱烘焙：**將麵糊慢慢倒入蛋糕杯，約2/3滿。用170℃烤20～30分鐘或烤至蛋糕表面呈金黃色。
7. **冷卻：**完成後移到散熱架上。等蛋糕完全冷卻後才裝飾。

裝飾步驟

1

2

3

1 **使用開縫星齒花嘴：**拿1個擠花袋，放入開縫星齒花嘴。將蛋白糖霜放入擠花袋，往下推實，紮緊袋口，在蛋糕表面擠上花紋。

2 **撒上糖珠：**淋上巧克力醬，再撒上彩色巧克力糖珠。

3 **放1塊鬆餅作裝飾：**放1塊鬆餅在蛋糕上作裝飾。

Belgium Waffle Cupcakes

INGREDIENTS

For Cupcake

90g unsalted butter
(at room temperature)
140g caster sugar
2 large eggs
180g self-raising flour
80ml skimmed milk
(at room temperature)
40g vegetable oil
1 teaspoon vanilla extract
6 wafer ice-cream cones
(can buy in supermarket)

For Decoration

Open Star Piping Tip
1 piping bag
1 portion of Meringue Frosting
(please refer to page 22)
rainbow chocolate sprinkles
chocolate sauce
(please refer to page 25)
waffle (put on the side for decoration)

CUPCAKE STEPS

1.Preheat the oven to 170 degrees Celsius.
2.Beat the butter and sugar until smooth.
3.Add in the egg and mix well.
4.Sift the self-raising flour. Add in the flour little by little to the batter.
5.Add in the milk little by little. Add in the vegetable oil. Add in the vanilla essence.
6.Pour into cupcake cups, 2/3 full. Bake for 20-30 minutes or until golden brown.
7.Transfer to cooling rack. Cool down completely before frosting.

DECORATION STEPS

1.Place the Open Star Piping Tip into a piping bag. Place the Meringue Frosting into the piping bag and pipe onto each cupcake.
2.Drizzle some chocolate sauce on top of meringue cream. Sprinkle rainbow chocolate sprinkles on top of cupcakes.
3.Place a small piece of waffle or cookie on the side of cupcake.

比利時焦糖餅乾蛋糕

Speculoos是比利時的傳統餅乾。餅乾裡面有杏仁粉，也有很多不同的香料。我第一次吃到這個餅乾，是在比利時的巧克力博物館。店員給我們1片沾了巧克力的Speculoos！味道好香，好特別！我想把這味道保留下來，就做了這個cupcake。如果大家有機會去比利時的話，除了買巧克力之外，還可以買罐Speculoos麵包醬，保證你一試難忘！

材料

焦糖餅乾

- 普通麵粉 250克
- 黑糖 185克
- 雞蛋 1顆
- 肉桂粉 1/4湯匙
- 無鹽奶油 125克
- 細磨杏仁粉 50克
- 混合香草粉 1/2湯匙

脆餅底

- 無鹽奶油 40克
- 比利時焦糖餅乾碎 80克

蛋 糕

- 無鹽奶油 45克
- 肉桂粉 1/2茶匙
- 雞蛋 1顆
- 自發粉 75克
- 杏仁粉 15克
- 脫脂牛奶 40毫升
- 植物油 40克
- 黑糖 70克
- 混合香草粉 1茶匙

裝 飾

- 擠花袋 1個
- 巧克力醬 1份（作法詳見25頁）
- 開縫星齒花嘴
- 杏仁碎 適量
- 比利時焦糖餅乾 6片

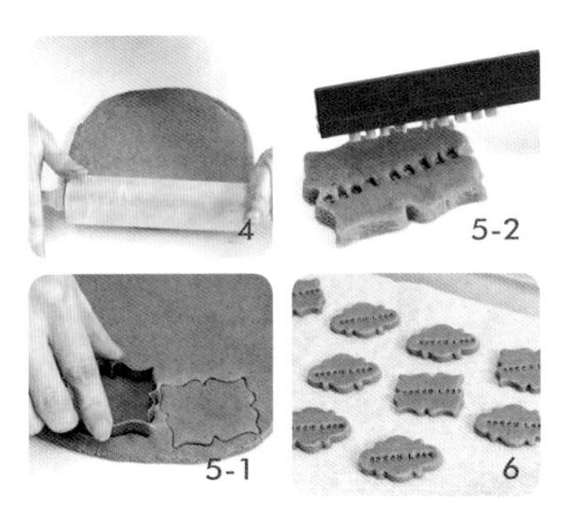

製作餅乾

1. **準備**：將烤箱預熱至160℃。奶油置於室溫中回軟。
2. **製作麵團**：混合所有材料，直到拌勻成麵團。
3. **冷藏**：用保鮮膜包好麵團，放入冰箱約35分鐘。
4. **壓平**：將麵團用擀麵棍壓平，推薄至約0.4～0.5公分厚。
5. **切模**：用餅乾模，切出不同形狀。
6. **入烤箱烘焙**：放入烤箱烤15分鐘或呈金黃色。

製作蛋糕

1. **製作蛋糕脆餅底層**：奶油融化後，加入餅乾碎，混勻。將餅乾碎放入蛋糕杯的底部，壓平至約0.4公分厚。
2. **預熱**：將烤箱預熱至170℃。
3. **打發奶油**：分3次加入砂糖，將奶油和糖混合打發，直到顏色變淺、呈軟滑狀。
4. **加入雞蛋**：加入雞蛋，攪拌均勻。
5. **篩入粉類**：將自發粉、杏仁粉、肉桂粉和香草粉混合過篩，然後慢慢加入雞蛋奶油溶液中。
6. **加入其他材料**：慢慢加入牛奶，再加入植物油。
7. **入烤箱烘焙**：將麵糊慢慢倒入蛋糕杯，約2/3滿。用170℃烤20～30分鐘或烤至蛋糕表面呈金黃色。
8. **冷卻**：完成後，轉移到散熱架上。等蛋糕完全冷卻後才裝飾。

裝飾步驟

1 **使用開縫星齒花嘴**：拿1個擠花袋，放入開縫星齒花嘴。將巧克力醬放入擠花袋，往下推實，紮緊袋口，在蛋糕表面擠上花紋。

2 **插上比利時焦糖餅乾**：插1片比利時焦糖餅乾在蛋糕表面作裝飾。再撒上杏仁碎。

Speculoos Cookies Cupcakes

INGREDIENTS

For Cupcake

250g all purpose flour
185g brown sugar
125g unsalted butter
(at room temperature)
1/4 tablespoon grounded cinnamon
1/2 tablespoon all spices
(powder of mixed ginger, nutmeg, etc)
1 large egg
50g almond powder

For Bottom layer

40g unsalted butter (melted)
80g speculoos cookies
(crumbs)

For Cupcake Layer

45g unsalted butter
(at room temperature)
70g brown sugar
1 large egg
75g self-raising flour
15g almond powder
1/2 teaspoon grounded cinnamon
1 teaspoon of all spices powder
40ml skimmed milk
(at room temperature)
20g vegetable oil

For Decoration

Open Star Piping Tip
1 piping bag
1 portion of Chocolate Ganache
(please refer to page 25)
6 speculoos cookies
chopped almond

COOKIES STEPS

1.Preheat the oven to 160 degrees celcius.
2.Mix all the ingredients together to form dough.
3.Put dough in the fridge for 35 minutes.
4.Roll out the dough 4-5mm thick.
5.Cut out shapes with a cookie cutter.
6.Bake for 15 minutes or until golden brown.

CUPCAKE STEPS

1.For bottom crust: Melt unsalted butter, mix in speculoos cookies into crumbs. Press firmly on bottom of each cupcake liners about 4mm thick.
2.For cupcake: Preheat the oven to 170 degrees Celsius.
3.Beat the butter and sugar until smooth.
4.Add in the egg and mix well.
5.Add the self-raising flour, cinnamon, all spice and almond powder mixture into batter.
6.Add in the milk little by little. And add in the vegetable oil.
7.Pour batter into cupcake cups on top of cookie crust, 2/3 full. Bake for 20-30 minutes or until golden brown.
8.Transfer to cooling rack. Cool down completely before frosting.

DECORATION STEPS

1.Place the Open Star Piping Tip into a piping bag. Place the Chocolate Ganache into the piping bag and pipe onto each cupcake.
2.Decorate with a speculoos cookie on top of each cupcake. Sprinkle with chopped almond.

白巧克力蛋糕

有些人喜歡吃白巧克力，我也是；當我吃了可可粉含量較多或濃度比較高的巧克力後，常會不停打噴嚏，所以更喜歡白巧克力。在這個cupcake裡，我加入了白巧克力粒，表面擠上白巧克力奶油霜，頂部再放上1塊白巧克力片作裝飾，賣相一流。在3月14日的白色情人節，男士們不妨考慮一下做這款cupcake送給你心儀的對象？

材料

蛋糕

□白巧克力 30克　□白巧克力粒 20克　□基本香草奶油蛋糕麵糊 1份（作法詳見19頁）

白巧克力奶油糖霜

□無鹽奶油 100克　□糖粉 200克　□脫脂牛奶 2～3茶匙

□白巧克力 70克　□高脂鮮奶油 1.5湯匙

裝飾

□擠花袋 1個　□白巧克力碎片　□多瓣花齒花嘴

□白巧克力片 6片　□食用彩色糖珠

製作蛋糕

1. **混合白巧克力和蛋糕糊：**隔水加熱白巧克力後，放入預先準備好的基本香草奶油蛋糕糊裡面。加入白巧克力粒。
2. **入烤箱烘焙：**將麵糊倒入蛋糕杯，約2/3滿。用170℃烤20～30分鐘或烤至蛋糕面呈金黃色。
3. **冷卻：**完成後，轉移到散熱架上。等蛋糕完全冷卻後才裝飾。

製作白巧克力奶油糖霜

1. **融化白巧克力：**隔水加熱白巧克力。
2. **打發奶油：**將室溫奶油打發至軟綿。將糖粉過篩，然後加入奶油內打發至順滑。
3. **加入其他材料：**慢慢加入脫脂牛奶。加入濃奶油拌勻。最後加入已經融化的白巧克力漿。
4. **繼續打發：**繼續打發至鬆軟、無粉粒狀態。

1

3

裝飾步驟

1. **使用多瓣花齒花嘴：**拿1個擠花袋，放入多瓣花齒花嘴。將白巧克力奶油糖霜放入擠花袋，往下推實，紮緊袋口，從蛋糕表面擠上花紋。
2. **撒上白巧克力碎片：**撒上白巧克力碎片和食用彩色糖珠作裝飾。最後插上1片白巧克力片。

White Chocolate Cupcake

INGREDIENTS

For Cupcake

1 portion of Basic Vanilla Cupcake Batter (please refer to page 19)
30g white chocolate (melted and fold in batter)
20g white chocolate chips

For White Chocolate Buttercream

100g unsalted butter (at room temperature)
200g icing sugar
2-3 teaspoon of skimmed milk
70g white chocolate (melted)
1.5 tablespoon double cream

For Decoration

Closed Star Piping Tip
1 piping bag
white chocolate flakes
6 white chocolate
edible sugar beads

CUPCAKE STEPS

1. Add the melted white chocolate into 1 portion of Basic Vanilla Cupcake Batter. Stir in the white chocolate-chips.
2. Pour into cupcake cups, 2/3 full. Bake for 20-30 minutes or until golden brown.
3. Transfer to cooling rack. Cool down completely before frosting.

BUTTERCREAM STEPS

1. Melt the chocolate in a small bowl. Beat the softened unsalted butter.
2. Sift the icing sugar in another bowl, and add in the mixture, little by little. Add in the milk, 1 teaspoon at a time.
3. Add in the double cream. Last add in the melted chocolate.
4. Beat until smooth and creamy.

DECORATION STEPS

1. Pipe the White Chocolate Buttercream on each cupcake using a Closed Star Piping Tip.
2. Sprinkle white chocolate flakes and sugar beads on top. Insert a piece of white chocolate bar on top.

黑巧克力蛋糕

代表比利時的味道怎麼能少了傳統的巧克力蛋糕！為了讓蛋糕有更加濃烈的巧克力味，我特別加入巧克力粒，而且蛋糕的頂部是選用巧克力和鮮奶油製成的巧克力醬，最後再用巧克力塊裝飾。這款cupcake絕對是巧克力迷的最愛！

材料

蛋糕

- 黑巧克力粒 15克
- 基本巧克力蛋糕麵糊 1份（作法詳見20頁）

裝飾

- 擠花袋 1個
- 巧克力醬 1份（作法詳見25頁）
- 多瓣花齒花嘴
- 無糖可可粉
- 黑巧克力碎
- 食用銀色糖珠

製作蛋糕

1 **放入黑巧克力：**將黑巧克力粒放入預先準備好的基本巧克力蛋糕糊裡面。

2 **入烤箱烘焙：**將麵糊倒入蛋糕杯，約2/3滿。用170℃烤20～30分鐘或烤至蛋糕面呈金黃色。

3 **冷卻：**完成後，轉移到散熱架上。等蛋糕完全冷卻後才裝飾。

裝飾步驟

1 **使用多瓣花齒花嘴：**拿1個擠花袋，放入多瓣花齒花嘴。將巧克力醬放入擠花袋，往下推實，紮緊袋口，在蛋糕表面擠上花紋。

2 **放上黑巧克力碎：**在巧克力醬上放上一些黑巧克力碎作裝飾。

3 **撒上無糖可可粉和銀色糖珠：**最後，撒上無糖可可粉和食用銀色糖珠。

1

2

3

Dark Chocolate Cupcakes

INGREDIENTS

For Cupcake

1 portion of Basic Chocolate Cupcake Batter (please refer to page 20)
15g dark chocolate chips

For Decoration

Closed Star Piping Tip
1 piping bag
1 portion of Chocolate Ganache (please refer to page 25)
dark chocolate flakes
cocoa powder
edible silver sugar beads

CUPCAKE STEPS

1. Add melted dark chocolate chips into 1 portion of Basic Chocolate Cupcake Batter.
2. Pour into cupcake cups, 2/3 full. Bake for 20-30 minutes or until golden brown.
3. Transfer to cooling rack. Cool down completely before frosting.

DECORATION STEPS

1. Place the Closed Star Piping Tip into a piping bag. Place the Chocolate Ganache into the piping bag and pipe onto each cupcake.
2. Decorate with chocolate flakes.
3. Sprinkle cocoa powder and edible silver sugar beads on top.

美國

UNITED STATES

美國是由50個洲和華盛頓哥倫比亞特區所組成，

因國土面積大，不同州區的天氣差異也很大。

因為這個原因，食材的選擇十分豐富。

很多人形容美國的飲食文化是以「comfort food」為主，

意思是指某些食物雖然製作簡單，

但卻能喚起家一般溫馨舒服的感覺。

不過不知道是否食物給了舒服感覺的原因，

人的胃口也大了，所以美國食物的份量都很大！

我特別喜歡美國的小吃，例如漢堡、甜甜圈，

感覺這裡的食物總是五彩繽紛的，

就好像當地人一樣樂觀風趣，充滿活力！

紅蘿蔔蛋糕

「紅蘿蔔蛋糕配上奶油乳酪糖霜」被譽為1970年代在美國最受歡迎的五大甜點之一。紅蘿蔔的糖分很高，所以最初被用來代替糖去做蛋糕，因為成本便宜很多。除了便宜外，紅蘿蔔還可以讓蛋糕更加濕潤、軟熟。再配上酸酸的奶油乳酪糖霜，可以中和蛋糕的甜，味道剛剛好！我還喜歡加入肉桂粉，味道更香，再來一些核桃，口感更佳！

材料

蛋糕

- 生紅蘿蔔 80克（切絲）
- 基本香草奶油蛋糕麵糊 1份（作法詳見19頁）
- 肉桂粉 1茶匙
- 核桃 20克（可選擇不加）

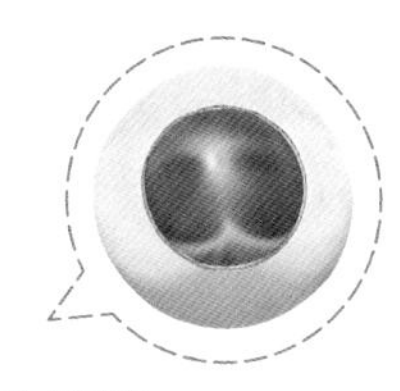

裝飾

- 擠花袋 1個
- 奶油乳酪糖霜 1份（作法詳見24頁）
- 圓口花嘴
- 大的橙色棉花糖 1粒
- 大的綠色棉花糖 1粒
- 糖做的白兔公仔（可選擇不加）

製作蛋糕

1 **混合材料：**將生紅蘿蔔、核桃、肉桂粉加入預先準備好的基本香草奶油蛋糕糊裡面。

2 **入烤箱烘焙：**將麵糊倒入蛋糕杯，約2/3滿。用170℃烤20～30分鐘或烤至蛋糕面呈金黃色。

3 **冷卻：**完成後，轉移到散熱架上。等蛋糕完全冷卻後才裝飾。

1-1

1-2

2

3

裝飾步驟

1 **使用圓口花嘴：**拿1個擠花袋，放入圓口花嘴。將奶油乳酪糖霜放入擠花袋，往下推實，紮緊袋口，在蛋糕表面擠上花紋。

2 **放上橙色棉花糖：**將橙色棉花糖剪開一半，放在奶油乳酪糖霜上面。

3 **放上綠色棉花糖：**將綠色棉花糖剪開一半，再剪成條狀，製成紅蘿蔔的頂部。

4 **放上白兔糖公仔（可選擇不加）：**最後，放1個糖做的白兔公仔作裝飾。

1

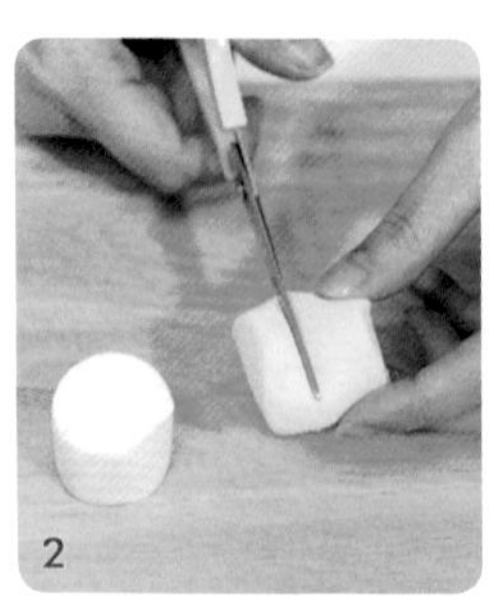
2

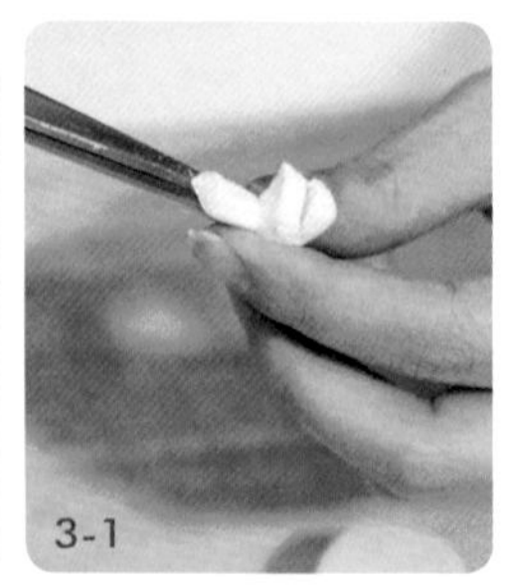
3-1

3-2

小技巧 利用食用色素增添色彩

如果棉花糖顏色不夠鮮艷，可噴上食用色素。

Carrot Cupcakes

INGREDIENTS

For Cupcake

1 portion of Basic Vanilla Cupcake Batter (please refer to page 19)
80g raw carrots (shredded)
20g chopped walnuts (optional)
1 teaspoon Grounded Cinnamon

For Decoration

Round Piping Tip
1 piping bag
1 portion of Cream Cheese (please refer to page 24) Frosting
1 large size orange fruity marshmallow
1 large size green fruity marshmallow
1 sugar rabbit (optional)

CUPCAKE STEPS

1. Add in grated raw carrots, walnuts (optional) and Grounded Cinnamon into 1 portion of Basic Vanilla Cupcake Batter.
2. Pour into cupcake cups, 2/3 full. Bake for 20-30 minutes or until golden brown
3. Transfer to cooling rack. Cool down completely before frosting.

DECORATION STEPS

1. Place the Round Star Piping Tip into a piping bag. Place the Cream Cheese Frosting into the piping bag and pipe onto each cupcake.
2. Cut the orange marshmallow into half and place on top of the frosting.
3. Cut the green marshmallow into half and cut some slits to make it like the stem of the carrot.
4. Place the sugar rabbit on the side (optional).

TIPS : **If the marshmallow's colour is not bright enough, you can also use food colorant.**

香蕉蛋糕

香蕉蛋糕是我的童年回憶！在我小時候，有位阿姨常常會烤香蕉蛋糕來我們家，她可以說是我烘焙興趣的啟蒙者！阿姨常常對我們說：用熟透香蕉做出來的香蕉蛋糕是最好吃的。而在我家中，一定會有香蕉出現，因為我爸媽都很喜歡吃這種水果。他們說香蕉有豐富的維他命B6、C、B2，對身體非常好。所以，下次有熟透的香蕉記得可留作做香蕉蛋糕哦！

材料

蛋糕

- 肉桂粉 1茶匙
- 香蕉 50克（已熟透，壓成泥醬）
- 基本香草奶油蛋糕麵糊 1份（作法詳見19頁）

裝飾

- 擠花袋 1個
- 鮮奶油 250毫升
- 開縫星齒花嘴
- 肉桂粉 適量
- 香蕉糖、新鮮香蕉或香蕉乾 6片

製作蛋糕

1 **混合香蕉、肉桂和奶油蛋糕糊**：將香蕉和肉桂粉加入預先準備好的基本香草奶油蛋糕麵糊裡面。

2 **入烤箱烘焙**：將麵糊倒入蛋糕杯，約2/3滿。用170℃烤20～30分鐘或烤至蛋糕面呈金黃色。

3 **冷卻**：完成後，轉移到散熱架上。等蛋糕完全冷卻後才裝飾。

小技巧 **最好不要在蛋糕中加入太多香蕉，否則會讓水分過多，難以發起。**

裝飾步驟

1 **打發鮮奶油**：將鮮奶油打至硬性發泡。

2 **使用開縫星齒花嘴**：拿1個擠花袋，放入開縫星齒花嘴。將鮮奶油放入擠花袋，往下推實，紮緊袋口，在蛋糕表面擠上花紋。

3 **用香蕉片、肉桂粉裝飾**：放上香蕉片或香蕉糖、香蕉乾，撒上肉桂粉。

Banana Cupcakes

INGREDIENTS

For Cupcake

1 portion of Basic Vanilla Cupcake Batter (please refer to page 19)
50g bananas (mashed)
1 teaspoon grounded cinnamon

For Decoration

Open Star Piping Tip
1 piping bag
250ml whipped cream
6 pieces of banana candies , fresh banana slices or dried bananas
grounded cinnamon

CUPCAKE STEPS

1.Add the bananas and grounded cinnamon into 1 portion of Vanilla Cupcakes Batter.
2.Pour into cupcake cups, 2/3 full. Bake for 20-30 minutes or until golden brown.
3.Transfer to cooling rack. Cool down completely before frosting.

DECORATION STEPS

1.Whip the cream to peak.
2.Place the Open Star Piping Tip into a piping bag. Place the whipped cream into the piping bag and pipe onto each cupcake.
3.Topped with fresh banana slices, dried bananas or banana candy. Sprinkled with grounded cinnamon on top.

TIPS : **Do not add too much banana to the batter, or the banana will become very soggy, as there is too much fluid.**

草莓起司蛋糕

美國的起司蛋糕通常都比較大，蛋糕的底大部分是用餅乾碎製成。這款草莓起司cupcake採用奶油蛋糕底，添加果醬和新鮮草莓，再配上香滑的奶油乳酪糖霜，口味更加大眾化，適合抗拒濃味起司的人食用。

材料

蛋糕

- □無鹽奶油 55克
- □白砂糖 50克
- □雞蛋 1顆
- □自發粉 100克
- □香草精 1/2茶匙
- □脫脂牛奶 40毫升
- □植物油 25克
- □草莓果醬 1/2湯匙
- □草莓 5粒（切成小粒果肉）

裝飾

- □奶油乳酪糖霜 1份（作法詳見24頁）
- □擠花袋 1個
- □多瓣花齒花嘴
- □草莓 3粒（切半）
- □消化餅乾碎
- □草莓果醬
- □粉紅色食用色素 1～2滴

製作蛋糕

1 **準備**：將烤箱預熱至170℃。奶油置於室溫中回軟。

2 **打發奶油**：打發奶油和糖，直到顏色變淺，呈軟滑狀。

3 **加入雞蛋**：加入雞蛋，攪拌均勻。

4 **篩入粉類**：將自發粉篩過，慢慢加入。

5 **加入其他材料**：依序加入脫脂牛奶、植物油、香草精。加入草莓果醬和草莓粒。

6 **入烤箱烘焙**：將麵糊倒入蛋糕杯，約2/3滿。用170℃烤20～30分鐘或烤至蛋糕面呈金黃色。

7 **冷卻**：完成後，轉移到散熱架上。等蛋糕完全冷卻後才裝飾。

裝飾步驟

1 **加入粉紅色素**：將粉紅色食用色素加入奶油乳酪糖霜裡。

2 **填入草莓果醬**：在蛋糕中間挖1小孔，用草莓果醬填滿。

3 **使用多瓣花齒花嘴**：拿1個擠花袋，放入多瓣花齒花嘴。將奶油乳酪糖霜放入擠花袋，往下推實，紮緊袋口，在蛋糕表面擠上花紋。

4 **用消化餅、鮮草莓作裝飾**：撒上消化餅乾碎。最後，放上鮮草莓作裝飾。

Strawberry Cheesecake Cupcakes

INGREDIENTS

For Cupcake
55g unsalted butter
(at room temperature)
50g caster sugar
1 large egg
100g self-raising flour
40ml skimmed milk
25ml vegetable oil
1/2 teaspoon vanilla essence
1/2 tablespoon strawberry jam
5 pcs strawberries
(cut into small pcs)

For Decoration
Closed Star Piping Tip
1 piping bag
1 portion of Cream Cheese Frosting
(please refer to page 24)
1-2 drops of pink food coloring
strawberry jam (as filling)
digestive biscuit crumbs
(for decoration)
3 strawberries (cut into halves)

CUPCAKE STEPS

1.Preheat the oven to 170 degrees Celsius.
2.Beat the butter and sugar until light yellow in color and smooth.
3.Add in the egg and mix well.
4.Sift the self-raising flour. Add in the flour little by little to the batter.
5.Add in the milk little by little. Add in the vegetable oil and the vanilla essence. Stir in the Strawberry Jam and strawberry bits.
6.Pour into cupcake cups, 2/3 full. Bake for 20-30 minutes or until golden brown.
7.Transfer to cooling rack. Cool down completely before frosting.

DECORATION STEPS

1.Add 1-2 drops of pink food coloring into the Cream Cheese Frosting.
2.Cut a hole in the middle of each cupcake. Fill Centre with strawberry jam.
3.Place the Closed Star Piping Tip into a piping bag. Place the Cream Cheese Frosting into the piping bag and pipe onto each cupcake.
4.Sprinkle with digestive biscuit crumbs. Decorate with half a fresh strawberry.

礁島萊姆派蛋糕

提到礁島萊姆派，一定要說它的起源。傳說是由一群漁民創造出來的。因為出海期間要帶一些可以儲存比較長時間，且含有豐富營養價值的食物，例如罐頭牛奶、煉乳、萊姆、雞蛋等等。而這些食物，就可以做到礁島萊姆派。不過因為船上不會有烤箱，所以傳統的做法是不用烤的。這次我就想到可以加一些蛋糕的元素去製成一個軟綿綿的礁島萊姆派蛋糕。

材料

脆餅底

□無鹽奶油 40克　□消化餅乾碎 80克

蛋 糕

□基本香草奶油蛋糕麵糊 1份（作法詳見19頁）□萊姆汁 2湯匙　□萊姆皮 1/2個

萊姆奶酪表層

□蛋黃 2顆　□煉乳 200毫升　□萊姆汁 1/4杯

裝 飾

□鮮奶油 250毫升　□萊姆皮碎　□新鮮萊姆片　□擠花袋 1個　□開縫星齒花嘴

製作蛋糕

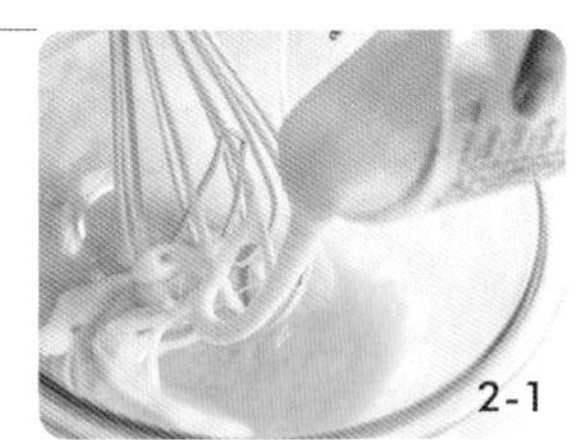
2-1

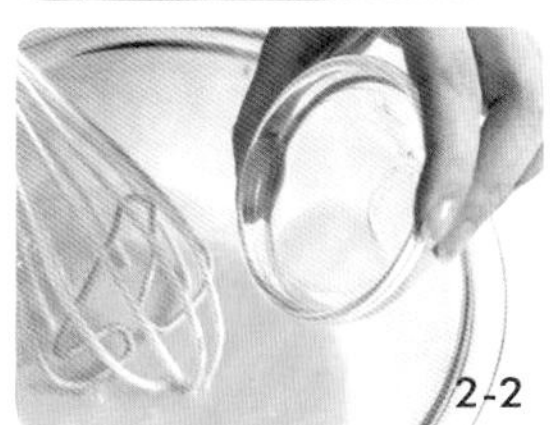
2-2

1 **製作脆餅底：**將奶油融化後，再與餅乾碎混合。將餅乾碎倒入蛋糕杯的底部，壓平至0.4～0.5公分厚。

2 **混合萊姆汁、萊姆皮和奶油蛋糕麵糊：**在預先準備好的基本香草奶油蛋糕麵糊裡加入萊姆汁和萊姆皮。

3 **入烤箱烘焙：**將麵糊慢慢倒入蛋糕杯，約 1/2滿。烘烤15～20分鐘，拿出來備用。

4 **製作萊姆奶酪表層：**將蛋黃、煉乳與萊姆汁混合，倒在蛋糕表面。用170℃烤20～30分鐘或烤至蛋糕表面呈金黃色。

5 **冷卻：**完成後，轉移到散熱架上。等蛋糕完全冷卻後才裝飾。

裝飾步驟

1 **打發鮮奶油：**將鮮奶油打至硬性發泡。

2 **使用開縫星齒花嘴：**拿1個擠花袋，放入開縫星齒花嘴。將鮮奶油放入擠花袋，往下推實，紮緊袋口，在蛋糕表面擠上花紋。

3 **用萊姆皮、萊姆片作裝飾：**撒上萊姆皮碎。最後，放1片新鮮萊姆片作裝飾。

Key Lime Pie Cupcakes

INGREDIENTS

For Bottom Layer
40g unsalted butter (melted)
80g digestive biscuits (crumbs)

For Cupcake
1 portion of Basic Vanilla Cupcake Batter (please refer to page 19)
2 tablespoon fresh lime juice
zest of 1/2 lime

For Key Lime Layer (top layer)
2 egg yolks
200ml condensed milk
1/4 cup fresh lime juice

For Decoration
Open Star Piping Tip
1 piping bag
250 ml whipping cream
lime zest for decoration
Fresh slices of lime or lime jelly candies

CUPCAKE STEPS

1. For the bottom crust: Melt the unsalted butter, add in digestive cookie crumbs, hand rub until all ingredients are well mixed together. Press cookie crumbs firmly at the bottom of each cupcake liner around 4-5mm thick.
2. For the cupcake: Add in the lime juice and lime zest into 1 portion of Basic Vanilla Cupcake Batter.
3. Pour into cupcake cups, 1/2 full. Bake for 15-20 minutes and take it out set aside.
4. For the Key Lime top Layer: Whisk the egg yolks, condensed milk and lime juice in a bowl until well blended. Pour on top of the cupcakes. Bake for another 10-15 minutes or until the top layer is sturdy.
5. Transfer to cooling rack. Cool down completely before frosting.

DECORATION STEPS

1. Place the Open Star Piping Tip into a piping bag.
2. Place the Whipped Cream into the piping bag and pipe onto each cupcake.
3. Sprinkle some lime zest on top of each cupcake. Decorate with a fresh slice of lime or lime jelly candies.

加拿大

CANADA

加拿大是我出生的地方，因此我對這裡有很多美好的回憶。

加拿大地方大，空氣好，風景美，

探訪當地時常會接觸到大自然，尤其是楓葉。

加拿大也是一個擁有多國文化的國家，有來自不同地方的人，

所以當地的美食也非常多元化。

不過，這個地方的食物通常比較直接，

意思是食物的賣相不花俏。烹調時，強調食材的新鮮，

烹調過程並不複雜。若要選擇代表加拿大的甜品，

我會選出那些在我成長中產生影響的，

希望透過這些特別的味道，讓大家對我的家鄉有更多了解。

楓樹糖漿蛋糕

在加拿大，楓樹無處不在！加拿大天氣很冷，楓樹在冬天的時候會製造糖。到了春天，這些糖就可以從楓樹汁液中提取出來。不過要經過加熱、蒸發水分等繁複工序後，楓樹糖漿才可食用。幾乎每個家庭都會有1瓶楓樹糖漿，用來搭配各種美食！我覺得它的味道很獨特，如果將它加入cupcake一起烘焙，一定會很香。

材料

蛋糕

- 無鹽奶油 55克
- 白砂糖 50克
- 雞蛋 1顆
- 自發粉 100克
- 植物油 25克
- 楓樹糖漿 20毫升
- 脫脂牛奶 40毫升

楓樹糖漿法式蛋白奶油霜

- 楓樹糖漿 2茶匙
- 食用色素 1～2滴（可選擇不加）
- 法式蛋白奶油霜 1份（作法詳見23頁）

裝飾

- 擠花袋 1個
- 葉形糖果或楓樹糖漿乾碎片
- 黑巧克力
- 圓口花嘴

製作蛋糕

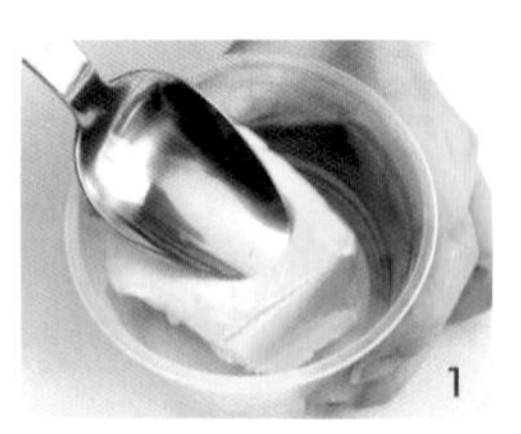

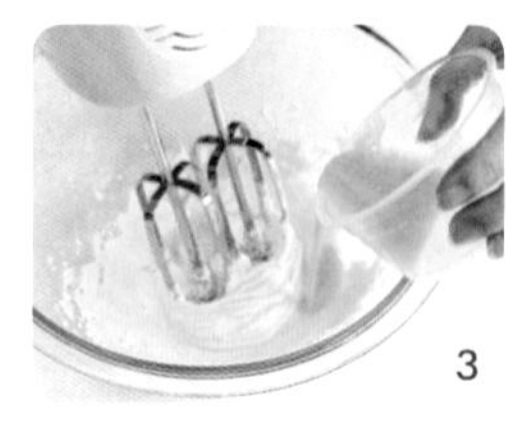

1 **準備：**將烤箱預熱至170℃。奶油在室溫中回軟。

2 **打發奶油：**加入白砂糖，和奶油一起打發，直到顏色變淺，呈軟滑狀。

3 **加入雞蛋：**加入打散的蛋液，混合攪拌，直到和奶油融合。

4 **篩入粉類：**將自發粉過篩，然後分3次加入。用矽膠刮刀以切拌的方式快速拌勻。

5 **加入其他材料：**加入脫脂牛奶、植物油、楓樹糖漿，攪拌均勻。

6 **入烤箱烘焙：**將麵糊慢慢倒入蛋糕杯，約2/3滿。用170℃烤20～30分鐘或烤至蛋糕表面呈金黃色。

7 **冷卻：**完成後移到散熱架上。等蛋糕完全冷卻後才裝飾。

製作楓樹糖漿法式蛋白奶油霜

1 **拌入楓樹糖漿：**將楓樹糖漿均勻拌入預先準備好的法式蛋白奶油霜中。

2 **加入食用色素：**如需調色，將食用色素滴入奶油霜中，再用抹刀混勻。

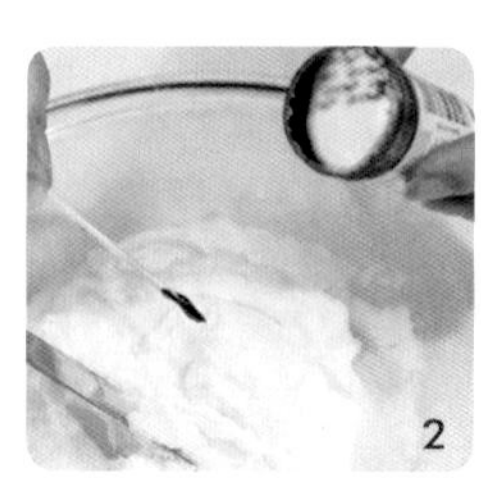

裝飾步驟

1. **使用圓口花嘴**：拿1個擠花袋，放入圓口花嘴。將楓樹糖漿法式蛋白奶油霜放入擠花袋，往下推實，紮緊袋口，在蛋糕表面擠上花紋。
2. **用巧克力畫出樹枝**：黑巧克力隔水加熱，裝入擠花袋，在烘焙紙上畫出樹枝。可先畫出簡單輪廓，再加粗。
3. **撒上糖果**：用葉形糖果或楓樹糖漿乾碎片作裝飾。
4. **放上奶油霜面**：將楓樹置於奶油霜上作裝飾。

1

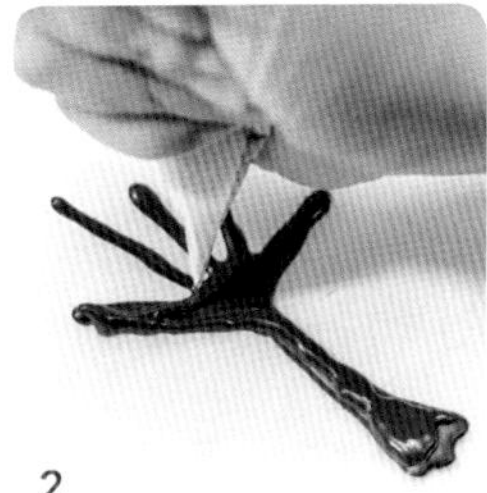
2

3

4

Maple Syrup Cupcakes

INGREDIENTS

For Cupcake

55g unsalted butter
(at room temperature)
50g caster sugar
1 large egg
100g self-raising flour
40ml skimmed milk
25ml vegetable oil
20ml maple syrup

For Maple Syrup French Meringue Buttercream

1 portion of French Meringue Buttercream
(please refer to page 23)
2 teaspoons Maple Syrup
1-2 drops of food coloring
(optional)

For Decoration

Round Piping Tip
1 piping bag
dark chocolate (for decoration)
leaf like candies or dried maple syrup bits (as decoration)

CUPCAKE STEPS

1. Preheat the oven to 170 degrees Celsius.
2. Beat the butter and sugar until smooth.
3. Add in the egg and mix well.
4. Sift the self-raising flour. Add in the flour little by little to the batter.
5. Add in the milk little by little. Add in the vegetable oil. Stir in the maple syrup
6. Pour into cupcake cups, 2/3 full. Bake for 20-30 minutes or until golden brown.
7. Transfer to cooling rack. Cool down completely before frosting.

MAPLE SYRUP FRENCH MERINGUE BUTTERCREAM STEPS

1. Stir in the Maple syrup into 1 portion of French Meringue Buttercream.
2. Add in 1-2 drops of food coloring (optional). Mix until smooth and creamy.

DECORATION STEPS

1. Place the Round Piping Tip into a piping bag. Place the Maple Syrup French Meringue Buttercream into the piping bag and pipe onto each cupcake.
2. Melt the dark chocolate. Use the melted chocolate and pipe out a tree with only branches onto baking paper. Start with a thin branch, let it dry and add extra layers to thicken it.
3. Decorate the chocolate branches with Leaf like candies/dried maple syrup bits as decoration.
4. Place on top of the buttercream.

加拿大
果仁巧克力蛋糕

Nanaimo bar是加拿大果仁巧克力蛋糕。名字是來自卑詩省一個叫做納奈莫（Nanaimo）的西岸城市，是加拿大最受歡迎的甜點之一。原本的食譜是不需要烘烤的，口感接近巧克力餅。亞洲人可能會覺得比較甜膩，所以我做了一個蛋糕版本。用原本的材料，加了一個蛋糕底，可以中和Nanaimo bar的甜度，同時又保留巧克力的濃香味道。

材料

脆餅底

- 無鹽奶油 40克
- 消化餅乾碎 80克

蛋 糕

- 椰絲 35克
- 基本巧克力蛋糕麵糊 1份（作法詳見20頁）
- 核桃或杏仁 10克

裝 飾

- 卡士達奶黃醬
- 椰絲 適量
- 核桃或杏仁
- 開縫星齒花嘴
- 卡通紙旗裝飾
- 擠花袋 1個
- 巧克力醬 1份（作法詳見25頁）

製作蛋糕

1 **製作脆餅底**：將融化的奶油與消化餅乾碎混合，再倒入蛋糕紙杯的底部，壓平至約0.4公分厚。

2 **混合蛋糕糊和其他材料**：預先準備好基本巧克力蛋糕麵糊，拌入椰絲、核桃或杏仁。

3 **入爐烘焙**：將麵糊慢慢倒入蛋糕紙杯，約2/3滿。用170℃烤20～30分鐘或烤至蛋糕表面呈金黃色。

4 **冷卻**：完成後，轉移到散熱架上。等蛋糕完全冷卻後才裝飾。

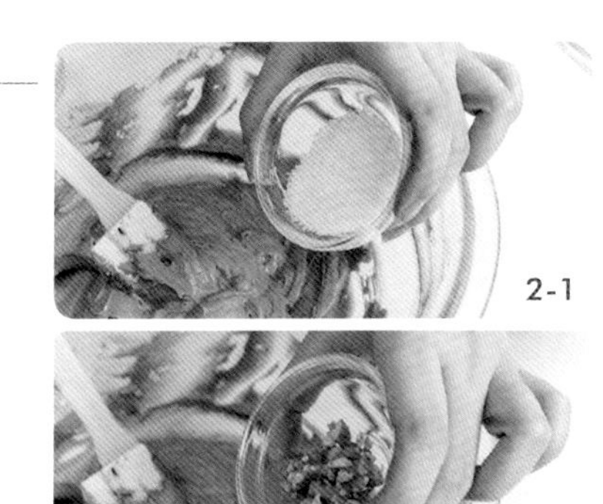

2-1

2-2

裝飾步驟

1 **填入卡士達奶黃醬**：在蛋糕中間挖1個小孔，填入卡士達奶黃醬。

2 **使用開縫星齒花嘴**：預備1個擠花袋，放入開縫星齒花嘴。然後放入巧克力醬，往下推實，紮緊袋口，在蛋糕表面擠上花紋。

3 **使用核桃、紙旗作裝飾**：在蛋糕上撒一些椰絲、核桃或杏仁，再插上印有卡通圖案的紙旗作裝飾。

Nanaimo Bar Cupcakes

INGREDIENTS

For Cupcake

Bottom layer
40g unsalted butter (at room temperature)
80g digestive biscuits (crumbs)

For Cupcake Layer

1 portion of Basic Chocolate Cupcake Batter (please refer to page 20)
10g almonds or walnuts
35g desiccated coconut

For Decoration

Open Star Piping Tip
1 piping bag
1 portion of Chocolate Ganache (please refer to page 25)
devon custard
desiccated coconut
walnuts/almonds
cartoon paper toppers

CUPCAKE STEPS

1. For the bottom crust: Melt the unsalted butter and add in the digestive biscuits. Press firmly at the bottom of each cupcake liners about 4mm thick.
2. For the cupcake: Fold in the coconut and walnuts/almonds into 1 portion of Basic Chocolate Cupcake Batter.
3. Pour into cupcake cups, 2/3 full. Bake for 20-30 minutes or until golden brown.
4. Transfer to cooling rack. Cool down completely before frosting.

DECORATION STEPS

1. Cut a hole in each cupcake. Fill centre with devon custard.
2. Place the Open Star Piping Tip into a piping bag. Place the Chocolate Ganache into the piping bag and pipe onto each cupcake.
3. Sprinkle desiccated coconut and walnuts on top. Decorate with cartoon paper toppers.

薄荷巧克力蛋糕

薄荷是很清新芳香的！在加拿大的時候，我弟弟最愛吃一款薄荷巧克力，但是現在已經停產，市面上也買不到。我們常常懷念這個巧克力的味道，所以我做了這款薄荷巧克力cupcake！這款cupcake由巧克力蛋糕底加上薄荷巧克力碎，再配上又香又滑的薄荷奶油糖霜組成。蛋糕香濃可口，讓我們可以重拾已經失去很久的味道。

材料

蛋糕

- 基本巧克力蛋糕麵糊 1份（作法詳見20頁）
- 薄荷香精 1/2茶匙
- 可烘焙薄荷巧克力粒 30克（可選擇不加）

薄荷奶油糖霜

- 基本奶油糖霜 1份（作法詳見21頁）
- 綠色食用色素 1～2滴
- 薄荷香精 1/2茶匙

裝飾

- 綠色裝飾彩砂糖 適量
- 擠花袋 1個
- 黑巧克力 適量
- 圓口花嘴
- 薄荷巧克力粒 適量
- 薄荷巧克力（可選擇不加）

製作蛋糕

1 **混合薄荷香精和巧克力麵糊：**將薄荷香精均勻拌入預先預備好的基本巧克力蛋糕麵糊裡。

2 **拌入巧克力粒：**加入可烘焙薄荷巧克力粒，拌勻。

3 **入烤箱烘焙：**將麵糊倒入蛋糕杯，約2/3滿。用170℃烤20～30分鐘或烤至蛋糕面呈金黃色。

4 **冷卻：**完成後，轉移到散熱架上。等蛋糕完全冷卻後才裝飾。

製作薄荷奶油糖霜

1 **拌入薄荷香精：**將薄荷香精拌勻入預先準備好的基本奶油糖霜裡。

2 **加入綠色色素：**加入綠色食用色素，拌勻。

3 **打發：**打發至鬆軟滑身。

裝飾步驟

1 **使用圓口花嘴：**拿1個擠花袋，放入圓口花嘴。將薄荷奶油糖霜放入擠花袋，往下推實，紮緊袋口，在蛋糕表面擠上花紋。

2 **用巧克力粒作裝飾：**在糖霜表面淋上隔水加熱的黑巧克力醬，撒一些薄荷巧克力粒、綠色裝飾彩砂糖或加1片薄荷巧克力作裝飾。

Mint Chocolate Cupcakes

INGREDIENTS

For Cupcake

1 portion of Basic Chocolate Cupcake Batter (please refer to page 20)
1/2 teaspoon peppermint essence
30g peppermint baking chocolate chips (optional)

For Peppermint Buttercream

1 portion of Basic Buttercream (please refer to page 21)
1/2 teaspoon peppermint essence
1-2 drops of green food coloring

For Decoration

Round Piping Tip
1 piping bag
peppermint chocolate chips
dark chocolate (melted)
green color sanded sugar
mint chocolate (optional)

CUPCAKE STEPS

1. Add in the peppermint essence to 1 portion of Basic Chocolate Cupcake Batter.
2. Add in the peppermint chocolate chips.
3. Pour into cupcake cups, 2/3 full. Bake for 20-30 minutes or until golden brown.
4. Transfer to cooling rack. Cool down completely before frosting.

PEPPERMINT BUTTERCREAM STEPS

1. Add in the peppermint essence to 1 portion of Basic Buttercream.
2. Add in 1-2 drops of green food coloring.
3. Beat until well blended and smooth.

DECORATION STEPS

1. Place the Round Piping Tip into a piping bag. Place the Peppermint Buttercream into the piping bag and pipe onto each cupcake.
2. Drizzle with melted dark chocolate, and sprinkle with peppermint chocolate chips, green color sanded sugar, or a piece of mint chocolate on top.

蘋果奶酥蛋糕

我在加拿大長大，對蘋果奶酥（金寶）情有獨鐘，每次見到菜單上有這道甜點，我都一定會點。我很喜歡暖暖的蘋果肉，帶著很香的肉桂味，再加上烤得脆脆的口感，實在是美味極了！最後千萬不能忘了冰淇淋這個「最佳拍檔」！這款蘋果奶酥cupcake就是升級版的蘋果奶酥，因為又多了一層軟綿綿的蛋糕口感。

材料

蛋糕

- 肉桂粉 1茶匙
- 青蘋果 1個（切粒）
- 基本香草奶油蛋糕麵糊 1份（作法詳見19頁）

奶酥

- 麵粉 50克
- 黑糖 30克
- 無鹽奶油 35克

製作蛋糕

1 **混合肉桂粉和蛋糕麵糊**：將肉桂粉加入預先準備好的基本香草奶油蛋糕麵糊裡面，拌勻。

2 **加入青蘋果粒**：青蘋果切粒，拌入麵糊裡。

3 **倒入蛋糕杯**：將麵糊慢慢倒入蛋糕杯，約2/3滿。

製作奶酥

1 **混合麵粉、黑糖、奶油**：將麵粉和黑糖混合。再加入切成細方塊的室溫奶油。

2 **搓拌**：用手將材料摩擦混拌，至麵包呈現碎粒狀。

3 **平鋪在蛋糕頂部**：將碎粒鋪在每個蛋糕的頂部。

4 **烘焙**：用170℃烤20～30分鐘至碎粒呈棕色，蛋糕表面呈金黃色。完成後，可配上冰淇淋。

1-1

1-2

2

3

Apple Crumble Cupcakes

INGREDIENTS

For Cupcake

1 portion of Basic Vanilla Cupcake Batter (please refer to page 19)
1 teaspoon grounded cinnamon
1 green apple (chopped)

For the crumble

50g plain flour
30g brown sugar
35g unsalted butter (at room temperature)

CUPCAKE STEPS

1. Add the grounded cinnamon to the Basic Vanilla Cupcake Batter.
2. Cut the apple into pieces and fold it into mixture.
3. Pour into cupcake cups, 2/3 full.

CRUMBLE TOPPINGS STEPS

1. Mix the flour and sugar. Cut the butter in cubes.
2. Rub in the butter using your fingers to mix into bread crumbs texture.
3. Sprinkle crumble on top of each cupcake.
4. Bake for 20-30 minutes until crumbs are brown and cupcake is golden brown. Served with ice cream (optional).

杯子蛋糕幸福上桌

糕體、糖霜到裝飾，輕鬆完成

作　　者　戚黛黛、蒙順意
編　　輯　黃馨慧
美術設計　周惠敏
校　　對　陳思穎

發 行 人　程安琪
總 策 畫　程顯灝
總 編 輯　呂增娣
主　　編　李瓊絲
編　　輯　鄭婷尹、陳思穎、邱昌昊、黃馨慧
美術主編　吳怡嫻
資深美編　劉錦堂
美術編輯　侯心苹
行銷總監　呂增慧
行銷企劃　謝儀方、吳孟蓉

發 行 部　侯莉莉
財 務 部　許麗娟
印　　務　許丁財
出 版 者　橘子文化事業有限公司

總 代 理　三友圖書有限公司
地　　址　106 台北市安和路 2 段 213 號 4 樓
電　　話　(02) 2377-4155
傳　　真　(02) 2377-4355
E－mail　service@sanyau.com.tw
郵政劃撥　05844889 三友圖書有限公司

總 經 銷　大和書報圖書股份有限公司
地　　址　新北市新莊區五工五路 2 號
電　　話　(02) 8990-2588
傳　　真　(02) 2299-7900

製　　版　興旺彩色印刷製版有限公司
印　　刷　鴻海科技印刷股份有限公司
初　　版　2016 年 1 月
定　　價　新臺幣 350 元
ＩＳＢＮ　978-986-364-081-3（平裝）

國家圖書館出版品預行編目（CIP）資料

杯子蛋糕幸福上桌：糕體、糖霜到裝飾，輕鬆完成 / 戚黛黛、蒙順意著. -- 初版. -- 臺北市：橘子文化, 2016.01
　面；　公分
ISBN 978-986-364-081-3（平裝）

1. 點心食譜

427.16　　104026566

本書繁體中文版由香港萬里機構・飲食天地出版社授權在臺灣地區出版發行